The Riddle of the Modern World

Also by Alan Macfarlane

WITCHCRAFT IN TUDOR AND STUART ENGLAND

THE FAMILY LIFE OF RALPH JOSSELIN

RESOURCES AND POPULATION: A Study of the Gurungs of Central Nepal

RECONSTRUCTING HISTORICAL COMMUNITIES (*with Sarah Harrison and Charles Jardine*)

THE ORIGINS OF ENGLISH INDIVIDUALISM

THE JUSTICE AND THE MARE'S ALE: Law and Disorder in Seventeenth-Century England (*with Sarah Harrison*)

A GUIDE TO ENGLISH HISTORICAL RECORDS

MARRIAGE AND LOVE IN ENGLAND, 1300–1840

THE CULTURE OF CAPITALISM

THE SAVAGE WARS OF PEACE: England, Japan and the Malthusian Trap

The Riddle of the Modern World

Of Liberty, Wealth and Equality

Alan Macfarlane
King's College
Cambridge

© Alan Macfarlane 2000

All rights reserved. No reproduction, copy or transmission of this publication may be made without written permission.

No paragraph of this publication may be reproduced, copied or transmitted save with written permission or in accordance with the provisions of the Copyright, Designs and Patents Act 1988, or under the terms of any licence permitting limited copying issued by the Copyright Licensing Agency, 90 Tottenham Court Road, London W1T 4LP.

Any person who does any unauthorised act in relation to this publication may be liable to criminal prosecution and civil claims for damages.

The author has asserted his right to be identified as the author of this work in accordance with the Copyright, Designs and Patents Act 1988.

First published in hardcover 2000

First published in paperback 2002 by
PALGRAVE
Houndmills, Basingstoke, Hampshire RG21 6XS and
175 Fifth Avenue, New York, N. Y. 10010
Companies and representatives throughout the world

PALGRAVE is the new global academic imprint of
St. Martin's Press LLC Scholarly and Reference Division and
Palgrave Publishers Ltd (formerly Macmillan Press Ltd).

ISBN 0–333–79270–X hardback (*outside North America*)
ISBN 0–312–23204–7 hardback (*in North America*)
ISBN 0–333–98450–1 paperback (*worldwide*)

This book is printed on paper suitable for recycling and made from fully managed and sustained forest sources.

A catalogue record for this book is available from the British Library.

The Library of Congress has cataloged the hardcover edition as follows:
Macfarlane, Alan.
　The riddle of the modern world : of liberty, wealth and equality / Alan Macfarlane.
　　p. cm.
　Includes bibliographical references and index.
　ISBN 0–312–23204–7
　1. Liberty. 2. Wealth. 3. Equality. 4. Montesquieu, Charles de Secondat, baron de, 1689–1755. 5. Smith, Adam, 1723-1790. 6. Tocqueville, Alexis de, 1805–1859. 7. Gellner, Ernest. I. Title.

JC585 .M43　2000
320'.01'—dc21

00–035260

10　9　8　7　6　5　4　3　2　1
11　10　09　08　07　06　05　04　03　02

Printed in Great Britain by Antony Rowe Ltd, Chippenham, Wiltshire

To the memory of Ernest Gellner (1925–95)

Contents

Acknowledgements viii
Note on References, Conventions and Measures x
Abbreviated Titles xii

1 The Riddle of the Modern World 1

I Liberty 11

2 Baron de Montesquieu's Life and Vision 13
3 Liberty and Despotism 29
4 The Defence of Liberty 47

II Wealth 71

5 Adam Smith's Life and Vision 73
6 Growth and Stasis 90
7 Of Wealth and Liberty 107
8 From Predation to Production 123

III Equality 151

9 Alexis de Tocqueville's Life and Vision 153
10 'America' as a Thought Experiment 167
11 How the Modern World Emerged 193
12 Liberty, Wealth and Equality 221

IV An Answer to the Riddle? 249

13 Ernest Gellner and the Conditions of the Exit 251
14 The Riddle Resolved? 269

Notes 295
Bibliography 321
Index 324

Acknowledgements

I would like to thank a number of people who have helped in various ways. These include Katherine Lang, Anne Muller, Caspian Richards and Dr David Sneath. John Davey read the typescript and was his usual encouraging and wise self. The University of Tokyo, and particularly Professor Takeo Funabiki, provided a sabbatical term during which, among other things, I rewrote sections of the text. The Research Centre at King's College, Cambridge, funded three seminars which were connected to the themes of this book. The participants in the seminars helped me clarify a number of ideas. Marilyn Strathern shielded me from administrative pressures and provided wise leadership. David Gellner read an early draft of my writing on his father, Ernest, and kindly gave me, on behalf of his family, permission to dedicate the book to him. Lily Blakely by her birth helped me to bring the final chapter to life.

Cherry Bryant read an earlier draft and made helpful suggestions for improvements. Iris Macfarlane read various drafts and made many useful comments. Cecilia Scurrah Ehrhart helped with checking the footnotes. Penny Lang typed and retyped the text, helped check the notes and aided in many other ways.

In particular I would like to thank two people. Gerry Martin for many stimulating ideas and conversations, for reading numerous drafts of the whole book and discussing it with me at length, and for moral and financial support through the Renaissance Trust. In many ways this is a collaborative work with him. Also I thank Hilda Martin for her friendship, encouragement and support. Sarah Harrison has, as always, given enormous help in every possible way, including several constructive readings of the text which helped to shorten it by a quarter. This book is likewise a collaborative work with her.

Finally I would like to acknowledge my debt to Ernest Gellner. I knew Ernest from 1968 to his death in November 1995 and his ideas have had an enormous effect on me. This book has grown out of various attempts to solve some of the riddles which Ernest had posed so well. In particular from the time of the publication of his *Plough, Sword and Book* (1987), and most brilliantly in *The Conditions of Liberty* (1996) I found that Ernest expressed more powerfully than anyone else now writing the Weberian sense of the astonishing nature of the 'escape' from *agraria*, against all the odds. He was always aware of the contingent, accidental nature of the emergence of our world. Thus he knew what the riddle was, even if his solution was as unsatisfactory to me as my earlier attempts were to him. It is a great sadness that he is not here to continue the conversation, but I dedicate this book with respect and affection to one of the last great Enlightenment thinkers.

Note on References, Conventions and Measures

Normally, much of the materials in a monograph would be one's own. This book, however, is a work in which I try to let four Enlightenment figures speak to us in their own words, hence the very large number of quotations from the authors considered. I hope that the authenticity of the many thoughts of the chosen authors will enrich the argument.

Spelling has not been modernized. American spelling (e.g. 'labor' for 'labour') has usually been changed to the English variant. Italics in quotations are in the original, unless otherwise indicated. Variant spellings in quotations have not been corrected.

Round brackets in quotations are those of the original author; my interpolations are in square brackets. In the quotations from Adam Smith, where angled brackets have been used by modern editors to show where words or letters have been added, I have reproduced these. Square brackets are used for my interpolations or expansions.

The footnote references give an abbreviated title and page number. The usual form is author, short title, volume number if there is one (in upper-case roman numerals), page number(s). The full title of the work referred to is given in the bibliography at the end of the book, where there is also a list of common abbreviations used in the footnotes.

Measures

A number of the quotations refer to English systems of measurement, some of which are now no longer in use.

Value: 4 farthings to a penny, 12 pennies (d) to a shilling (s), 20 shillings to a pound (£). One pound in the seventeenth century was worth about 40 times its present value (in 1997).

Weight: 16 ounces to a pound, 14 pounds to a stone, 8 stone to one hundredweight (cwt) and 20 hundredweight to a ton. (one pound (lb) equals approx. 0.454 kg.)

Liquid volume: 2 pints to a quart, 4 quarts to a gallon. (approx 1.75 pints to 1 litre)

Distance: 12 inches to a foot, 3 feet to a yard, 1760 yards to a mile. (approx 39.4 inches to 1 metre)
Area: an acre. (approx 2.47 acres to 1 hectare)

Abbreviated Titles

Titles of works by the four major thinkers discussed in this book are abbreviated as follows.

Ernest Gellner (1920–95)

Devil	*The Devil in Modern Philosophy*, 1974
Legitimation	*Legitimation and Belief*, Cambridge, 1979
Muslim	*Muslim Society*, Cambridge, 1981
Spectacles	*Spectacles and Predicaments: Essays in Social Theory*, Cambridge, 1981
Nations	*Nations and Nationalism*, Oxford, 1983
Plough	*Plough, Sword and Book: The Structure of Human History*, 1988
'Interview':	'An Interview with Ernest Gellner' by John Davis in *Current Anthropology*, vol. 32, no. 1, Feb. 1991
Anthropology	*Anthropology and Politics: Revolutions in the Sacred Grove*, Oxford, 1995
Culture	*Culture, Identity and Politics*, Cambridge, 1995
Liberty	*Conditions of Liberty, Civil Society and Its Rivals*, 1995
Gellner	*The Social Philosophy of Ernest Gellner*, ed. John Hall and Ian Jarvie, Amsterdam, 1996

Baron de Montesquieu (1689–1755)

Persian Letters	*Persian Letters* (1721), tr. John Davidson, no date (*c.*1891)
Considerations	*Considerations on the Decline of the Romans* (1734), tr. David Lowenthal, New York, 1965
Spirit	*The Spirit of the Laws* (1748), tr. Thomas Nugent, 2 vols in 1, New York 1975
Pensées	*Montesquieu par lui-même*, by Jean Starobinski, Paris, 1959

Abbreviated Titles xiii

Adam Smith (1723-90)

Moral	*The Theory of Moral Sentiments* (1759), George Bell, 1907
Wealth	*An Inquiry into the Nature and Causes of the Wealth of Nations* (1775), 2 vols in 1, Univ. of Chicago Press paperback edn., 1976
Philosophical	*Essays on Philosophical Subjects*, Indianapolis, 1982
Jurisprudence	*Lectures on Jurisprudence*, Indianapolis, 1982
Rhetoric	*Lectures of Rhetoric and Belles Lettres*, Indianapolis, 1985
Correspondence	*Correspondence of Adam Smith*, Indianapolis, 1987

Alexis de Tocqueville (1805-59)

Journey to America	*Journey to America*, tr. George Lawrence, 1959
Recollections	*The Recollections of Alexis de Tocqueville* (1893), ed. J.P. Mayer, tr. Alexander Teixeira de Mattos, 1948
Journey	*Journeys to England and Ireland*, ed. J.P. Mayer, tr. George Lawrence and K.P. Mayer, New York, 1968
Memoir	*Memoir, Letters, and Remains of Alexis de Tocqueville*, 2 vols, Cambridge 1961
Democracy	*Democracy in America* (1835, 1840), 2 vols, tr. George Lawrence, 1968
Ancien	*L'Ancien Regime (1856)*, tr. M.W. Patterson, Oxford, 1956
'Notes'	Notes to *The Ancien Régime and the French Revolution*, tr. Stuart Gilbert, 1966
Letters	*Selected Letters on Politics and Society*, 1985
European Revolution	*'The European Revolution' and Correspondence with Gobineau*, tr. John Lukacs, Connecticut, 1959

1
The Riddle of the Modern World

> Know then thyself, presume not God to scan,
> The proper study of mankind is man.
> Placed on this isthmus of a middle state,
> A being darkly wise, and rudely great:
> With too much knowledge for the sceptic side,
> With too much weakness for the stoic's pride.
> He hangs between; in doubt to act, or rest;
> In doubt to deem himself a god, or beast;
> In doubt his mind or body to prefer;
> Born but to die, and reasoning but to err;
> Alike in ignorance, his reason such,
> Whether he thinks too little, or too much:
> Chaos of thought and passion, all confused;
> Still by himself abused, or disabused;
> Created half to rise, and half to fall;
> Great lord of all things, yet a prey to all;
> Sole judge of truth, in endless error hurl'd;
> The glory, jest, and riddle of the world!
>
> (Alexander Pope, Essay on Man, Epistle II)

In these famous lines written in 1732 Alexander Pope attempted to summarize the quintessential nature of human beings and their predicament. Pope did not know that the sense of contradiction which he felt was particularly intense in that very era because he was living on one of the great fault-lines of history. Mankind was just starting to witness spectacular changes which would alter all the parameters of human life, first in England, and later for all humans and all other species on this planet.

As we stand at the start of the third millennium, it is timely to consider what those changes were which have transformed our world in the last 300 years. We need to understand something about that mixture of political, economic and ideological factors to which we give rough labels such as 'industrialism', 'capitalism', 'democracy', 'modernity', and of the older concepts of liberty, wealth and equality which lie behind them. As these forces sweep across much of Asia and eastern Europe, transforming the lives of billions of people, as they only recently transformed the lives of millions in 'the West', it is even more urgent to understand their nature causes and consequences.

Yet this is a surprisingly difficult task, for the emergence of our modern world and its very nature is a mystery. We are very confused as to how it came about, or what it is that surrounds us so closely that we can scarcely see it. In order to gain some distance, this book will examine two inter-connected aspects of what I have termed 'the riddle of our world'. One aim is to attempt to discover some of the forces which have made the development of human civilizations in the past so slow and painful. The second is the opposite: what has enabled the economic 'progress' of parts of humankind to be so amazingly rapid during the last 300 years?

The first riddle is posed by the fact that we might have expected that since the rise of agrarian civilizations about five thousand years ago, mankind's economic progress would have been much faster than it has been. This expectation would be based on some characteristics that make human beings different from other animals on this planet.[1] It could be argued that there are two fundamental characteristics of *Homo sapiens sapiens* which in degree at least differentiate us from all other extant species. The first is that we, as humans, have the physiological and cultural potential for intellectual activity which permits us to generate knowledge of the natural world and of ourselves as part of that world. We are above all a knowledge-generating and knowledge-using species. The second is that we have the ability to use such knowledge to modify the world around us, most notably to produce artifacts which we mobilize as tools for survival.

The fact that life on earth, over the millennia of recorded history, has led to the ultimate outcome of a species which can manipulate, almost atom by atom, its own genome appears to be related most directly to these two features. We are the species which generates its own resources, by manipulation of the matter of which our planet is composed, using energy derived from our star, the sun. We do this, always, by the application of knowledge. In theory human beings are capable of using a very powerful and cumulatively effective feedback loop: the innovation of knowledge, leading to innovation of functional artifacts (including agricultural produce as an artifact) which leads to their multiplication and back to the innovation of knowledge. Each time the cycle is repeated the process becomes more powerful. There should thus be cumulative growth in knowledge and artifacts.

Yet when we consider the 'progress' of mankind over the thousands of years up to the start of the eighteenth century it is clear that the various political, economic and ideological systems within which man's creativity is embedded have forced a departure from this expected growth of knowledge and control of nature. On almost every level, there seemed to have been a steady worsening of the condition of the vast majority of those who lived in the great agrarian civilizations which dominated the globe. While gross economic wealth had increased and the earth sustained more and more people, while the technologies of thought and power gave mankind hitherto undreamt of control over the environment, the polarization between the lucky few and the misery of the masses had grown.

Narrowing the time frame, if an outside observer had been able to monitor the developments in human societies over the 500 years up to about 1700, he or she would not have been optimistic about the future of mankind. In the most populated area, China, some 200 million human beings had achieved a high level of sophistication and organization, but seemed to have reached an equilibrium within which there was little further technological, social or economic development. Indeed, many contemporary observers believed that the oldest and mightiest civilization on the planet was gradually losing much of its distinctive science and civilization. In the second most populous civilization, India, the great days of the Moghul emperors were coming to an end in petty rivalries. Famine and intermittent warfare spread as the European powers closed in to dismember the remains of the glories of Akbar and Shah Jahan. In the once scientifically, culturally and militarily pre-eminent civilization of Islam, learning, wealth and tolerance were in retreat as the glories of previous centuries were overlaid. In the populous Empire of Japan, the dynamic period of the early Tokugawa shoguns was over. Rapid population growth had filled the small area of cultivable land, and the first large famine in modern Japanese history was soon to occur. There was little innovation or growth and again an equilibrium seems to have been reached. The great empires of south America, Aztecs and Incas, had been destroyed and were but memories. Sub-Saharan and northern Africa were increasingly undermined by western incursions, the slave trade and epidemic disease. Much of eastern Europe had relapsed into a second serfdom and though Russia was vibrant under Peter the Great, the country, like much of central Asia, was still subject to massive famine and Mongol incursions. Western and southern Europe also seemed to have hit invisible barriers. The Mediterranean lands were suffering from increasing disease and the economies and populations were either static or declining. The great days of the Italian states, Spain and Portugal seemed to be over, and southern Germany had been devastated by the Thirty Years War. North-western Europe seemed to be the only area in the world where there was still some dynamic development, but even that was patchy. The once proud Hanseatic league had much declined. Sweden was crippled by the adventures of Gustavus Adolphus. It is true that Holland had reached the position of the richest nation in the world, but there were signs that its economy had stopped growing, and it was increasingly harassed by envious neighbours.

France was the most powerful nation in Europe, yet it suffered periodic famines, high mortality and the negative consequences of Louis XIV's expansionist dreams. Many of its most productive workers had been expelled when religious toleration ended with the Revocation of the Edict of Nantes in 1685. England was rich and powerful, but it had narrowly avoided being conquered in 1715 by a small army of Highland Scots, and the death rates from endemic and epidemic disease had started to rise in the later seventeenth century. North America was still a rural backwater where the destruction of its native population was continuing apace. Thus the world, with its roughly 500 million inhabitants, seemed to have reached the limit to its potential to support human life. Half a century before Malthus the prospects looked grim. Mankind seemed to be caught on a treadmill. Thus the first aspect of the riddle concerns the obstacles or traps which had hitherto halted the growth of all the great agrarian civilizations.

*

The second part of the riddle concerns what has happened since the eighteenth century. When we look back, 300 years later, from the end of the twentieth century, a mere speck of time in the history of man on earth, the tendencies briefly summarized above have almost all been partially reversed. Since this is a familiar story, I shall only very briefly summarize the totally unpredictable and surprising transformation. War has continued with us, but as a cause of human death it has relatively declined in importance. The trend seems to have started in the West in the early eighteenth century and in Asia from the nineteenth, though there have been some notable reversals. Large-scale famine disappeared in western Europe in the eighteenth century, with the tragic exception of Ireland. It disappeared in most of Asia after the 1960s. Disease mortality started to drop in parts of Europe from the eighteenth century, then in the later nineteenth century the causes of most epidemic diseases were discovered. Although there is now a resurgence of malaria and other diseases, and of new diseases such as AIDS, most epidemic and endemic diseases were under control in Europe by the later nineteenth century and in Asia by the 1960s.

The industrial revolution has freed many from long hours of physical work. In the West this happened from the later nineteenth century, and it is now starting to happen in many parts of Asia. Indeed, the major problem for many in the more affluent nations now is too little work, not too much. Machines driven by fossil fuels have increasingly taken the strain off the human back, arms and legs. Computers are replacing the human brain.

In terms of birth status, the famous declaration of the rights of man which proclaimed the certain truth of the fact that man is born equal and free was only one landmark in the reversal in the premise of inequality, whose latest expression is the end of apartheid. Slavery was formally abolished in the first half of the nineteenth century, then serfdom. Caste has been more obdurate and class has replaced the old system of estates. Yet whatever the continued inequalities, there is now a widespread belief in the universal equality of humans at birth. In terms of economic differentials, there has been a massive levelling. The very rich are still present, and there has recently been a tendency for the gap between rich and poor to open up again. At a wider view, however, there is no longer a vast gap between the 1–5 per cent who have 1000 times the income of the average of the rest. There is a more gradual gradient of wealth, with the majority of the population, at least in the West and increasingly in Asia, living at a level of affluence undreamt of even two centuries ago.

In terms of political power, the gradual death of absolutist regimes has gathered pace and 'democracy' has spread. First in parts of north-western Europe and America, then in India, Japan and southern Europe, then in south America, and now dramatically in eastern Europe and South Africa. Even in China there have been rumblings. Highly centralized and unified power seems to be on the retreat. Its last famous representatives, Stalin, Hitler, Mussolini, Mao and Pol Pot, have been only weakly replaced by the now 'amoral supermen', the communications managers. The ideal, if not the practice, of democracy is widely espoused and has become the norm, rather than the exception.[2]

There has also been an opening and levelling of knowledge. The closed worlds of priesthood and literati have begun to wither before mass education, the rapid spread of printing and other communications technology, the growth of scepticism and tolerance. The inquisition has been formally suspended and the thought police pensioned off in many parts of the world. The ordinary educated citizen may not have specialist knowledge in many fields, but the division of knowledge is no longer between the one who has keys to all that is known and the nineteen who are totally excluded. The emancipation of women in many parts of the world has been partially achieved, though huge inequalities remain. Foot-binding and *sutee* have been abolished, and though bride-burning and female circumcision and purdah still continue, there is little doubt that an impressive shift in power has occurred. Few, in the West at least, would openly subscribe to the view of women as a sinful, inferior creature, born to be the slave of man.

The overall material situation is well summarized by Mokyr. 'The riches of the post-industrial society have meant longer and healthier lives, liberation from the pains of hunger, from the fears of infant mortality, from the unrelenting deprivation that were the part of all but a very few in preindustrial society.'[3] Thus the second part of the riddle, put bluntly, is: why have humans done so well over the last 300 years?

It is easy enough to specify what was needed to effect this change. Basically, almost all the trends of the last 2000 years had to be reversed. Warfare had to be contained, famine abolished and disease minimized. The resulting lowered mortality rates had to be balanced by controlled fertility. Growing wealth had to be distributed more widely and the societies had to avoid the tendency towards rigid social stratification. The burden of work had to be lifted by the use of improved technology and the fruits of increasing knowledge shared through widespread education and literacy. The monopoly of the priesthood had to be curtailed and the domain of magic and supernatural causation minimized. The political structure had to develop in a way which avoided the build-up of absolutism, but also its opposite, fragmentation, and there had to be a widening participation in power.

This is easy enough to specify. The amazing fact is that these reversals, so apparently unlikely, did occur. So enormous were the changes that we tend to call the central ones 'revolutions'. New technologies were developed which allowed mankind to supplement current energy from animals and plants with the vast stored resources in coal and oil. A new organization of labour was developed based on factories, which allowed for the production of cheap goods for a mass market. This revolution was in turn based on a new scientific method, which encouraged the systematic exploration of the natural world.

These two revolutions provided the means for a massive increase in world population. This was a population which, first in the West and increasingly now in the rest of the world, has gone through a 'demographic revolution', at last escaping from the Malthusian checks of war, famine and disease and replacing perennial high and uncontrolled mortality and fertility with their opposite. This in its turn has resulted in the massive growth of cities. Accompanying these changes there has been a transformation in the relations of production. The normal division of agrarian societies into the four estates of warriors and rulers, clergy and literati, merchants and artisans, and peasants was replaced by a rather more fluid class structure of owners of capital and workers.

At the political level there was a move from rule by the rich or well born to rule by the people. This new political system was based on a novel premise, that all men were born equal. Linked to this was a movement from a status-based world, where people only had a meaning as members of a larger group, to one based on the belief that each human being was a separate individual, free, independent, with innate rights and duties. Finally there was a growth of rationality, that is to say the partial expulsion of magic, the elimination of miracles, the ever-closer link between means and ends, the acceptance of universal truths and a reason that is not context-bound. It is the cumulative effects of these revolutions that has created the modern world as we know it. How they occurred is still a mystery.

*

Alexis de Tocqueville noted that it is extremely difficult, after the event, to know how past changes occur. This is because 'great successful revolutions, by effecting the disappearance of the causes which brought them about, by their very success, become themselves incomprehensible'.[4] In many ways, this is our problem in understanding what has happened during the last 300 years. The very success of science, technology and democracy have made the causes of that success difficult to understand. We know *that* it happened, and we are easily led into the belief that it *had* to happen. Now that it appears to be happening in many parts of the globe, even the formidable doubts of someone like Max Weber seem irrelevant. Thus our greatest difficulty is that we know what has happened and hence the riddles themselves have disappeared. When we look back from our vantage point at the end of the twentieth century, all of this transformation looks inevitable, perhaps the working out of some 'Spirit of History', as Hegel and, more recently, Fukuyama have argued.[5] It is only when we survey the total trajectory of human history and pause for a moment at about 1700 that we see how extraordinary, unexpected and enormous the 'Great Transformation' has been.

*

Clearly there are innumerable ways in which one might approach these giant problems. In this book I have done so by considering the ideas of four major thinkers. Each of them asked the following questions: 'What is the nature of the new world that is coming into being, why has it come about, and what are its consequences?' Each gave a brilliant answer, deriving from their own method and context. Those chosen are the Baron de Montesquieu (1689–1755), Adam Smith (1723–90), Alexis de Tocqueville (1805–59) and Ernest Gellner (1925–95). I have also briefly considered the ideas of some of their contemporaries, especially Alexander Pope, Bernard Mandeville and David Hume. Just as an anthropologist tries to get inside the skin of another society by using the eyes and ears of key informants, so it seems reasonable to see whether we can reach the essence of the formative changes if we look into the minds and hearts of a few of those who experienced events most intensely and tried to understand them from within their own life experience.

Another reason for trying to understand the thinkers I have chosen is that they are very much part of the air we breathe. They not only reflected their ages, they shaped them, not only at the disciplinary level, but more deeply. The American and French Revolutions were shaped by Montesquieu's thought; the industrial and capitalist revolutions by Adam Smith. It has been said that we are all the slaves of dead philosophers. In order to free ourselves it is necessary to understand what they said and what part of our world they influenced. Of course there are many possible informants, and readers may rightly wonder why I have chosen these four and not others. By confining myself, as I shall do, to the period since 1715, I have ruled out many interesting earlier observers who could have given us clues – Montaigne, Bacon, Machiavelli, Hobbes, Locke to name only a few of the better-known. Yet, as we shall see, many of the most fruitful ideas of these earlier thinkers were absorbed into the work of our four writers.

The particular virtue of starting with Montesquieu in the second quarter of the eighteenth century is that, as we have already begun to see, this is precisely the point at which two things began to become apparent. The first was that in certain parts of Europe, particularly the north–west, the material conditions of life seemed to be beginning to improve. The second was that at the world level, certain parts of western Europe were starting visibly to 'progress' faster than Asia. Montesquieu stood on the threshold, before the industrial and urban revolutions had taken hold. Adam Smith worked as these revolutions were just starting. A biographical account of each of them will show that with their broad comparative approach, which encompasses what was known about India, China, Japan and elsewhere, as well as Europe, they provided deeply insightful accounts of the structural dynamics of agrarian civilization. They provide for us a dissection of the natural tendencies, ceilings, traps and mechanisms which we have already glimpsed in action over the last 5000 years. They sensed that something strange was happening in north-western Europe, but could not quite grasp what it was. Consequently they came to the conclusion that the 'escape' from agrarian misery was ultimately impossible. In the end they were pessimists. Liberty could increase for a while, but in the end it would collapse into despotism. Opulence could grow – for a while

– but quite soon within an agrarian system the limits to growth would be reached.

In order to pursue these questions further in time I have selected one further thinker whose work addressed the fundamental problems of the relations between liberty, equality, and wealth and who thus continued the search which the Enlightenment thinkers had begun. He is Alexis de Tocqueville, whose broad comparative approach focused on three of the major civilizations of the West, namely France, England and America. Tocqueville provides many insights into the nature of the new world that was in the process of being born, and particularly the form and implications of equality.

The insights of these three great thinkers into the nature of mankind, the contradictions of modernity and the necessary conditions for a humane civilization are not merely of historical interest. Much of the world as we know it has been based on their answers to the riddle of the emergence of the modern world. The issues they raised concerning the conflicts and contradictions which are built into all social systems, but which are particularly striking in the new type of civilization which emerged principally from the eighteenth century, are still at the heart of our search for order and meaning. What this search is, and how important it is, is finally illuminated by taking the work of one contemporary philosophic historian, Ernest Gellner, who has addressed much of his life to these broad Enlightenment questions. This book concludes with an assessment of his theories and a final concluding chapter which synthesizes the unusual answer which our authors have given to the riddle of the origin and nature of our world.

I Liberty

2
Baron de Montesquieu's Life and Vision

Charles-Louis de Secondat, Baron de la Brède and Montesquieu, was born at the chateau of La Brède, near Bordeaux, on 18 January 1689. His ancestors were soldiers and magistrates and he himself trained to be a lawyer. He was educated at the College of the Oratorians near Paris from the age of 11, and at 16 returned to the University of Bordeaux to continue his study of law. In 1708, at age 19, he was admitted to the bar as an *avocat au Parlement*. Further studies in Paris, probably as a law clerk, between 1709 and 1713 took place before his father's death in 1713 (his mother had died in 1696). He married Jeanne Lartigue in 1715; in 1716 his uncle died, so he took the name Montesquieu formally and became *président à mortier* at Bordeaux. In 1721 his *Les Lettres Persanes* were published anonymously in Holland and were a great success; he became a member of various salons in Paris. In 1726 he sold his office of president in Bordeaux and in effect retired from legal life.

During the years 1728–31 he travelled to Hungary, Italy, Austria, Germany and England, spending much the longest period in England. In 1734 *Les Considérations sur les causes de la grandeur des Romains et de leur décadences* was published in Holland. In 1735, aged 46, he explicitly began his work on *The Spirit of the Laws* which would fill the rest of his life. In 1748, some 13 years later, the *Spirit* was published in Geneva, again anonymously. In 1752 the book was placed on the Catholic Church's Index of Forbidden Books. Montesquieu died in Paris on 10 February 1755, aged 66.

In considering the context of his work, one powerful influence was a tension between the estate-owning aristocrat and the world of commerce. Not only was his mother 'shrewd in business', but Montesquieu's upbringing on the edge of the great international port of Bordeaux gave him an acquaintance with the world of trade and manufacture. Bordeaux was 'unusually cosmopolitan because the wine trade brought a lot of foreign businessmen to the city'.[1] It was for Montesquieu what Glasgow was to be for Adam Smith, a potent reminder of a wider international world, and in both cases the orientation was towards the west – to the expanding wealth of the Americas. Althusser captures some of this when he writes that Montesquieu's breadth of vision, his attempt to write a work that took as its object 'the entire history of all the men who have ever lived', was related to a revolution in the knowledge of the world which was occurring as a result of explorations in the ages succeeding Columbus. Bordeaux was perfectly placed to receive news of these distant lands. 'It is the age of the discovery of the Earth, of the great explorations opening up to Europe the knowledge and the exploitation of the Indies East and West and of Africa. Travellers brought back in their coffers spices and gold, and in their memories the tales of customs and institutions which overthrew all the received truths.'[2] This had laid the foundations for Montaigne's great relativist speculations. Two centuries on, Montesquieu had even richer data to draw on, for the new knowledge of India and particularly China and Japan was beginning to flow in as the great trading networks expanded over the world.

Not only did his position thrust a new knowledge upon him, but it brought home to him the power of commerce. We shall see that one of his major contributions was partly to overcome the normal aristocratic antipathy to business and production. The influence of his mother, of running the La Brède estate, and of commercial Bordeaux helped create that interest and insight. The tensions of his upbringing and background in Gascony were made all the stronger by his life's experience. One of the most important of these was the changing political and social world of France over his lifetime. Montesquieu was born and brought up in the shadow of Louis XIV, probably the greatest of the absolutist rulers in European history, who had reigned for 72 years from 1643 to 1715. Thus Montesquieu was already 26 when Louis died. Louis's avowed aim was to spread his rule over the whole of Europe. Thus Montesquieu saw the immense strength of the most powerful state of Europe all round him and he himself, as a judge who put people to torture, was an instrument of that power. Parallel and equally obvious was the power of the Roman Catholic Church, with its inquisition and censor. That Montesquieu had to publish his three major works outside

France, and two of them anonymously, is just one indication of the twin threats to liberty.

Thus he grew up in the hierarchical, all-encompassing, world of the *ancien régime* where politics and religion were joined and control was paramount. And then, to provide the shock of contrast, of another possibility, two things happened. The first was the death of Louis XIV which liberated France. The change is well described by Sorel, and by Saint-Simon, whom he quotes.

> Louis XIV had just departed. His declining years resembled a gloomy and majestic sunset. Contemporaries did not stop to admire the twilight of a great reign; they were glad to be set free. No one regretted the king; he had too strictly imposed on all Frenchmen 'that dependence which subjected all'. 'The provinces,' says Saint-Simon, 'rallying from despair at their ruin and annihilation, breathed free and trembled for joy. The higher courts and the whole magisterial caste had been reduced to insignificance by edicts and appeals; now the former hoped to make a figure, the latter to be exempt from royal intermeddling. The people, ruined, crushed, and desperate, thanked Heaven with scandalous openness for a deliverance touching the reality of which their eagerness admitted no doubts.'[3]

There followed a period of relative freedom, excitement and openness. Montesquieu was exhilarated, and the *Lettres Persanes*, published six years later, were a product of this more liberal and open world. Yet it was, within France, only a relative liberation. Those very letters had to be published in Holland and the French censor never formally allowed their entry to France. The power of the Catholic Church was little diminished. France was still a modified *ancien régime*. Montesquieu could read about alternative, more open, systems and his interest in early Greek and particularly Roman civilization gave him models. Yet what he needed in order to prove that an alternative, open world was indeed possible, was a large scale living example. This example was provided by his visit to England.

Montesquieu arrived in England on 23rd October 1729. He stayed for nearly two years and closely studied the political and social system. When he arrived, he wrote, 'I am here in a country which hardly resembles the rest of Europe.'[4] The English were a 'free people'; as opposed to other nations 'this nation is passionately fond of liberty', 'every individual is independent', 'with regard to religion, as in this state every subject has a free will, and must consequently be . . . conducted by the light of his own mind . . . by which means the number of sects is increased.'[5] Summarizing the interconnections he suggested that the English have 'progressed the farthest of all peoples of the world in three important things: in piety, in commerce, and in freedom'.[6]

Montesquieu himself summarized his debt to the various countries which he visited; Germany was made to travel in, Italy to sojourn in, France to live in, and England to think in.[7] As Collins puts it in his useful description of Montesquieu's visit to England, the trip

> transformed the author of the *Persian Letters* and of the *Temple de Gnide* into the author of the *Considérations sur la Grandeur et Décadence des Romains* and of the *Esprit des Lois*. The study of our constitution, of our politics, of our laws, of our temper and idiosyncrasies, of our social system, of our customs, manners, and habits, furnished him with material which was indispensable to the production of his great work.[8]

Montesquieu found many faults with England which are revealed in the few scraps that remain from his time there. For instance, in relation to religion, he thought the English had gone too far. '"There is," he writes in his *Notes*, "no religion in England; in the Houses of Parliament prayers are never attended by more than four or five members, except on great occasions. If one speaks of religion, every one laughs." The very phrase "an article of faith" provokes ridicule.' Referring to the committee which had recently been appointed to inquire into the state of religion, he says that it was regarded with contempt. In France he himself passed as having too little religion, in England as having too much; and yet, he grimly adds, 'there is no nation that has more need of religion than the English, for those who are not afraid to hang themselves ought to be afraid of being damned.'[9] Likewise, he found it a very lonely, isolated, place, the people cold and unfriendly. 'When I am in France I make friends with everyone; in England I make friends with no one.'[10]

Yet it was an excellent place to think in, or even, as a kind of living model, to think with. He could use this experience and his many discussions and reading to construct an alternative to the *ancien régime*. Here was something old, continuous, yet very new. A powerful, commercial, tolerant nation that challenged all his own assumptions. It provided the actual, worked example of a system that could stand up against despotism. He realized its weaknesses, but he also saw its growing strength, both as a practice and an ideal. As Collins puts it,

> It was here that he saw illustrated, as it were in epitome and with all the emphasis of glaring contrast, the virtues, the vices, the potentialities of good, the potentialities of evil, inherent in monarchy, in aristocracy, in the power of the people. It was here that he perceived and understood what liberty meant, intellectually, morally, politically, socially. He saw it in its ugliness, he saw it in its beauty.[11]

England was a place to think in, so both there and when he returned to France with this model before him, 'Patiently, soberly, without prejudice, without heat, he investigated, analysed, sifted, balanced; and on the conclusions that he drew were founded most of the generalisations which have made him immortal.'[12] 'It was in England that the ideas to be developed in both these masterpieces [*Considérations* and *Spirit of the Laws*] took a definite form, in England that they found stimulus and inspiration, from England that they drew nutriment.'[13]

Montesquieu had now the two experiences that seem to be necessary for deep analysis of the foundations of a civilization. On the one hand there is the personal knowledge of a rapid and dramatic change within one's own country and environment: this was provided by the pre and post Louis XIV world. The second was the shock of placing one's own society and its assumptions against those of another – in this case *ancien régime* France against the strangely different England.

*

In attempting to understand himself, to understand a changing France, to make sense of the broad topics which he wished to study, Montesquieu had to come to terms with a number of theoretical problems which have continued to face the social sciences. The central problem concerns the nature of cause and effect, within which we can include single and multiple causation, general and particular causes and the nature of change through time. This was an area that he found so important, yet difficult, that among his unpublished archives is an *Essai sur les causes*, which was written 'to clarify one of his most difficult problems, that of the relationship between physical and moral causes'.[14]

It seems that Montesquieu, like any true historian faced with the evidence, was ambivalent and contradictory on the question of chance and necessity. On the one hand he recognized that apparently accidental and random small events could have immense consequences and change the course of history. Thus in his *Pensées* he wrote 'all these great movements only happened because of some unplanned, unforeseen action. Lucrecia's death caused Tarquin's fall. Brutus's act of executing his sons established liberty. Seeing Virginia slain by her father caused the fall of the Ten.'[15] These are small, personal, almost accidental events, yet 'From the public's unsuspected and therefore unpredictable reaction flows a new social order'.[16]

On the other hand, he is much better known for his view that history is mainly governed by general causes, in which individual humans are just epiphenomena. One example also occurs in his *Pensées*. He distinguishes between the general causes which were bound to lead to a Reformation and the accident of Martin Luther. 'Martin Luther is credited with the Reformation. But it had to happen. If it had not been Luther, it would have been someone else. The arts and sciences coming from Greece had already opened eyes to abuses. Such a cause had to produce some effect. A proof of this: the councils of Constance and Basel had introduced a kind of reformation.'[17] In a similar way he wrote, 'It was not the affair of Pultowa that ruined Charles. Had he not been destroyed at that place, he would have been in another. The casualties of fortune are easily repaired; but who can be guarded against events that incessantly arise from the nature of things?'[18] A more famous example comes in his account of the rise and decline of the Roman Empire.

> It is not chance that rules the world. Ask the Romans, who had a continuous sequence of successes when they were guided by a certain plan, and an uninterrupted sequence of reverses when they followed another. There are general causes, moral and physical, which act in every monarchy, elevating it, maintaining it, or hurling it to the ground. All accidents are controlled by these causes. And if the chance of one battle – that is, a particular

cause – has brought a state to ruin, some general cause made it necessary for that state to perish from a single battle. In a word, the main trend draws with it all particular accidents.[19]

Towards the end of this passage he almost brings the two together. A single battle may topple a state, it is the proximate cause, but it only acts in this way as a result of deeper background causes. His whole discussion of this matter, as Shackleton shows, was related to the work of contemporaries, in particular Vico and Doria. In fact the first half of the passage above is directly inspired by *La Vita civile* by Doria.[20] As Shackleton says, this is not to disparage Montesquieu, who carried into history 'the distinction between the First Cause and occasional causes'.[21]

Montesquieu's ultimate aim was to understand the cause or causes of things; why some societies suffered from despotism, why northern Europe was growing richer, why the world's population seemed to have declined and why the Roman Empire had collapsed. Like all historians, his views changed over time, and he swung from being an almost naive geographical and climatic determinist, as Collingwood thought him to be,[22] to being an almost idealist thinker who believed that moral and intellectual causes swayed history. His thought is a bundle of contradictions.[23]

If we look at all of his work, we see that he ends up with a balanced approach to causation. The 'spirit of the laws' as constituted by a number of interacting causes, physical and moral, is well summarized in a passage from Montesquieu quoted by Sorel. The spirit of a people's law

> must have relations to the physical characteristics of a country, to the climate, – frigid, torrid, or temperate, – to the nature of the land, its situation, its extent, to the people's mode of life; . . . they must have relations to the degree of liberty that the constitution can admit of; to the religion of the inhabitants, their inclinations, their wealth, their number, their commerce, their morals, their manners. Finally, they have relations one to another, to their origin, to the aim of the legislator, to the order of things under which they were established. They must be considered from all these points of view, and I undertake so to consider them in this work. I shall examine all these relations; together they form what I have called the *Spirit of the Laws*.[24]

20 The Riddle of the Modern World

This is an excellent brief outline for the social sciences and it does not show any kind of naive physical or moral determinism. In seeking such a balance, Montesquieu was a true ancestor of Durkheim, Weber and twentieth-century anthropology, even if, in practice, his book, written in fits and starts, occasionally falls away from this balanced view.

Montesquieu's attitude towards time and progress is complex, and this is reflected in an apparent disagreement among those who have commented on his theories. On the one hand a number of authors have argued that he had little sense of historical change and cumulative progress. For example, J.B. Bury in *The Idea of Progress* states that 'Montesquieu was not among the apostles of the idea of Progress. It never secured any hold upon his mind'. He finds this odd, for 'he had grown up in the same intellectual climate in which that idea was produced'. Yet he failed to grasp it. 'Whatever be the value of the idea of Progress, we may agree with Comte that, if Montesquieu had grasped it, he would have produced a more striking work.'[25] Richter writes that 'Montesquieu did not believe in the theory of progress; his philosophy of history has been described as "pessimism in moderation".'[26] Likewise Shklar states that 'He did not believe in cumulative progress, and his sense of the most recent past was one of radical discontinuity rather than continuous development. It seemed to him that the expansion of Europe after the discovery of America had made it so wealthy and powerful that it was wholly unlike anything that had ever existed in the past.'[27]

On the other hand Shackleton, his noted biographer and commentator, argues that Montesquieu did have a sense of progressive time. In an article first published in 1949, Shackleton makes a detailed study of the *esprit général* as a central organizing concept in Montesquieu's thought. He concludes that this 'gives the lie... to those who deny to Montesquieu any evolutionary sense and shows that he has a clear and straightforward theory of progress'.[28] In his later biography of Montesquieu Shackleton takes up the theme again. He shows that while Montesquieu's theory is 'more tentative and more empirical' than Turgot's, 'he has enunciated, not less than Turgot, a theory of progress'.[29] This is the progress which human kind has made from a world ruled by physical causes, to one governed by moral and legal forces.

The two views of Montesquieu in fact reflect two strands in his thought. He believed that societies and mankind evolved, changed, grew more complex, altered, just as a tree grows and branches. But he also refused to believe that moral progress, happiness, security, freedom and all the valuable things in life necessarily increased. As he looked at the collapse of Rome, at the repeated invasions of the Mongols, at the cruelties and despotism he thought ruled much of the world, at the short duration of open societies, he felt no great confidence in the future. His was an interesting and exactly balanced mixture of a cyclical and linear view. He lived on the cusp of the times, just at the point when the world seemed about to escape from that nightmare of the *ancien régime* which he dreaded. Montesquieu stands exactly on the bridge between the *ancien régime* and something entirely unpredictable and surprising. His greatness lies in the fact that in his ambivalent reflections on his times he sensed, often only implicitly, what was just below the horizon.

Montesquieu wished to understand the whole of world history and the whole of his current world. The *Spirit of the Laws* 'has for its object the laws, customs, and various usages of all peoples'.[30] In order to do this he developed a series of methods which laid the basis for the social and historical sciences. One of these was his comparative methodology, which aimed to compare not only the different parts of Europe, but Europe with the Islamic societies of the Middle East and even Europe with China and Japan. In attempting this huge comparison Montesquieu was fortunate, for it was just at the time that a massive influx of new information about other civilizations began to arrive in Europe. The greatest account of Japanese civilization in a foreign language, Engelbert Kaempfer's *History of Japan*, was published in three volumes in 1727 and Du Halde's *Empire of China*, an encyclopedic compilation of early missionary accounts of China, published in 1735. Montesquieu also used the collections of accounts of early trade missions to China. The *Spirit of the Laws* is thus notable in that it is the first great comparative survey of world civilizations. Montaigne had attempted something similar one and a half centuries earlier, also writing from near the port of Bordeaux. But Montaigne's sources on the Far East were exiguous. The Chinese dimension was intellectually important for Montesquieu because it made it possible to see the whole of Europe as a system or civilization with an innate dynamism in contrast to the stasis which had apparently overtaken China.

Durkheim believed that Montesquieu's implicit use of the comparative method was a central feature of his work.[31] Richter suggests that Montesquieu was somewhat more explicitly aware of what he was doing, citing him to suggest that 'Comparison, the single most valuable capacity of the human mind, is particularly useful when applied to human collectivities'.³² Thus he believes that Montesquieu 'made comparison the central problem of political sociology and thus directed the forms of inquiry away from Europe to all the societies known, however imperfectly, to man'.³³ This is a task which is fundamental to anthropology as well. It also has the effect of putting one's own society into doubt: 'Montesquieu argued that we can understand political and social phenomena only when we can stipulate some arrangement alternative to that in question.'³⁴

Yet in order to compare across societies one has to engage in a form of classificatory activity which is both difficult and unusual. This led Montesquieu into a second methodological innovation. He is widely credited with introducing 'ideal type' analysis, that is the setting up of simplified models against which reality can be tested, benchmarks, so to speak. For instance, he set up such models of the three forms of political organization: republics, monarchies and despotism. This had been foreshadowed by Aristotle and Machiavelli and others, but Montesquieu took the analysis much further not only by showing the forms but also by analysing the socio-geographical conditions within which each occurred. He is well aware that actual cases do not correspond to the ideal type, and thus there are frequent references to the fact that England as a place may not conform to the 'ideal-type' England he has created, and likewise that Chinese despotism in practice is not like Chinese despotism in its theoretical construction. Each European nation had deviated from his ideal-type picture of monarchy. What the ideal types did, however, was to allow Montesquieu, and later Weber and others, to engage in fruitful comparative research.³⁵

Although he did not believe in teleological evolution or inevitable progress towards a predestined goal, Montesquieu was both interested and well versed in history. There are several strands to his work which are historical and two can be mentioned here. One is his historical interest in Roman civilization, shown both in his *Considerations* and in a number of chapters of *Spirit*. Although his treatment of Rome has been criticized, what Roman history allowed Montesquieu to do was to watch the process of historical change over a long period, to see the whole of a civilization's growth, greatness and decay and to analyse the reasons for the latter. The central message was that all civilizations contain within themselves the seeds of their own destruction. Rome collapsed through internal corruption, and particularly because the balance of power at the centre became skewed. A model of how this had happened and would happen to all successful empires was of great use to Montesquieu.

His second interest was in the origins of modern France, which took him into many years of work on the historical sources for the Dark Ages. He wrote at length on the early foundations of liberty in the customs of the peoples who swept across Europe at the collapse of Roman civilization, and then traced the early stages of the evolution of feudalism in Europe. His curiosity 'attracted him towards those mysterious forests whence issued along with the Germans, his alleged ancestors, the elements of political liberty'. This research was lengthy and laborious.' 'His toil was severe, his investigations slow and painful. "I seem," he remarked, "all at sea, and in a shoreless sea. All these cold, dry, tasteless, and difficult writings must be read, must be devoured...."'[36] But it was true historical research and certainly imbued with an idea of difference and change over time. It was not all on one flat, asynchronous, level plain.

A third important part of his vision is what we might term a 'structural' or 'relational' approach to history and society. In the *Persian Letters*, Montesquieu foreshadowed a structural definition of his central concept, *laws*, when he wrote: 'Justice is a true relation existing between two things; a relation which is always the same, whoever contemplates it, whether it be God, or an angel, or lastly, man himself.'[37] Thus justice does not lie in a thing, but in a relation between things: perhaps this is why *chose* (the French for 'thing') comes up so much in his conversation. He was talking of those 'relations of relations' which is at the heart of structural thinking. The very first sentence of *The Spirit of the Laws*, the key definition, which has puzzled and often upset so many readers, proclaims this same structural approach. 'Laws, in their most general signification, are the necessary relations arising from the nature of things. In this sense all beings have their laws: the Deity His laws, the material world its laws, the intelligences superior to man their laws, the beasts their laws, man his laws.'[38] Notice here that 'relations' cover all aspects of life. Everything is relational. In case the reader has not grasped the point, he reiterates two paragraphs later that 'laws are the relations subsisting between it and different beings, and the relations of these to one another'.[39] 'It' in this case is the intelligence or reasoning power of human beings.

As Sorel noted, this is a very general, almost mathematical concept. Montesquieu 'rightly intimates that this definition is very wide. It is so wide that it eludes analysis and reaches out toward infinity. It is an algebraic formula, applying to all real quantities and expressing none of them exactly. It is rigorously true of mathematical and natural laws; its application to political and civil laws is only remote and rather indistinct.'[40] As Durkheim noted, 'he informs the reader that he intends to deal with social science in an almost mathematical way'.[41]

It greatly puzzled many, especially lawyers, that Montesquieu should start with such an unusual, novel and abstract definition. If, however, we consider the later development of structural thinking, particularly in the French tradition through Durkheim and Mauss, Saussure and Lévi-Strauss, we can see what Montesquieu was doing. He was able to see the relations between power, wealth and belief, or politics, economics and religion, in a way which was hitherto impossible. This ability to connect or relate the hitherto unconnected is well put by Fletcher.

> The supreme value of the method is that it permits the observer to see particular historical data in an entirely new setting, and hence to perceive relationships between things which, through their separation in time or space, must otherwise have remained unrelated. By applying the method over a sufficiently wide area of experience, and comparing similar 'relationships' in whatever time or place each to each, the general 'spirit' governing them all can be revealed.[42]

Montesquieu's concept of social structures, of a kind of machine in which there are relations of parts to each other and to the whole, also feeds into the functionalism and structural functionalism which came to dominate the social sciences between the 1880s and 1940s. This is noted by Shklar when she writes that for him 'Society is a system of norms which are related to each other and can be understood historically and as functioning to maintain the social whole'.[43] This is straight early twentieth-century functionalism.[44]

Most of those who have studied Montesquieu's writings, including the unpublished notebooks and his library, agree that like all scientists he mixed induction and deduction. An example is given by Shackleton. 'The development of Montesquieu's thoughts in relation to climate shows itself as being clearly inductive. Starting with an examination of the specific problem of Roman air, enlarging his ideas by reading, by observation, and by experiment, he arrives in the end at his general theory of climatic influence.'[45] This is fairly characteristic, yet it conceals a deductive phase, as Fletcher notes.

> It is true that he usually starts from particular 'facts' and then extracts from them a general principle. By a sufficiently vast correlation of such 'facts' and of the particular principles which emerge from them, he is enabled by his inductive process to reach those very broad generalisations which he calls the "principles" of the leading types of government. But thereafter the method becomes deductive. 'When I have discovered my principles' he says in his Preface, 'all that I am searching for comes to me.' And again: 'I propose my principles, and I look to see if the particular cases fit with them.'[46]

Montesquieu was even more explicit elsewhere: 'You do not invent a system after having read history; but you begin with the system and then you seek the proof. There are so many facts in a long history, people have thought so differently, the beginnings are usually so obscure that you always find enough to support all kinds of reactions.'[47]

Durkheim echoes Montesquieu's own assessment. Montesquieu

> does not begin by marshalling all the facts relevant to the subject, by setting them forth so that they can be examined and evaluated objectively. For the most part, he attempts by pure deduction to prove the idea he has already formed. He shows that it is implicit in the nature or, if you will, in the *essence* of man, society, trade, religion, in short, in the definition of the things in question. Only then does he set forth the facts which in his opinion confirm his hypothesis.[48]

Thus

> If we examine Montesquieu's own demonstrations, it is easy to see that they are essentially deductive. True, he usually substantiates his conclusions by observation, but this entire part of his argumentation is very weak. The facts he borrows from history are set forth briefly and summarily, and he goes to little pains to establish their veracity, even when they are controversial.[49]

Thus in brief, Montesquieu 'instead of using deduction to interpret what has been proved by experiment, he uses experiment to illustrate the conclusions of deduction'.[50]

Montesquieu's vast canvas, his attempt to cover the whole of the world and the whole of human history, and to connect all the different aspects of life, as well as his primarily deductive method, led to many inaccuracies and sins of omission and commission which commentators have pointed to. For instance Shackleton summarizes some of the criticism of his accuracy and historical methods in relation to Roman history.[51] Sorel wrote:

> As Montesquieu had failed to have recourse to archaeology and textual criticism in his study of primitive Rome, so now, in like manner, he failed to utilize anthropology in his study of primitive society. Why could he not have read Buffon? The 'Seventh Epoch of Nature' would have explained primitive humanity and the origin of customs to him in a very simple way.[52]

He could be accused of finding what he wanted to find. An example is in his treatment of East Asia. His account of Japan, almost exclusively stressing the harshness of Japanese law and punishment, gives a distorted picture of that civilization.[53] He has read Kaempfer's great three-volume work on Japan, and yet abstracts only what is relevant to his argument. His treatment of China is particularly interesting. As Richter points out, Montesquieu was the first to make the new discoveries in China, particularly Du Halde's compilations, available to a wider audience and his depiction was enormously influential. Yet it is skewed towards a picture of absolutist despotism.[54]

In fact, as a number of writers have pointed out, Montesquieu was very puzzled by what he read about China and confused by the contrary depictions they gave.[55] Thus his account is somewhat contradictory, trying to make sense of conflicting images of a despotic and a benign government, of a nation ruled by fear or by kindness.[56] In his writing we can hear Montesquieu arguing with himself. Should he modify his model of Asiatic despotism, originally based on the model provided by Machiavelli and others of the Ottoman Empire, or reject the data from some of his Chinese sources? In the end the model wins, though it is a little modified. This is, as in all of his accounts, both his strength and his weakness. Without the over-simplification we would not have his amazing insights into the structural causes of the decline of Rome, the roots of feudal systems in Europe, and the essence of English political institutions. But in all of these, as in his treatment of the Orient, we must guard against the distortions of a powerful mind fitting data to a pre-conceived framework.

The danger becomes particularly great as the immense scope of his undertaking reveals itself and tiredness, growing blindness and more and more data overwhelm him. 'In my view, my work grows in proportion to my diminishing strength. I have however, eighteen nearly finished books and eight which need arranging. If I was not mad about it, I would not write a single line. But what sorrows me is to see the beautiful things which I could do if I had eyes.'[57] Montesquieu questioned his own wisdom in taking on the task. 'I have laboured for twenty years on this work, and I no longer know if I have been bold or if I have been rash, if I had been overwhelmed by the size of my subject or if I had been sustained by its majesty.'[58]

Montesquieu himself, quoted by Sorel, perceived the difficulty of an ever-expanding project and of growing weariness.

> So long as he worked upon the earlier books he was all joy and ardour. 'My great work is going forward with giant strides,' he wrote in 1744 to the Abbé Guasco. Then was the time when 'all he sought came to him of itself'. But little by little masses of facts accumulated at the outlets and blocked them up. He forces the facts. 'Everything yields to my principles,' he wrote toward the last; but he does not see 'particular cases smoothly conforming to them', as formerly. He makes an effort, canvasses the texts, arrays analogies, heaps up, but he no longer welds together. He settles himself doggedly to the task; he grows fatigued. 'I am reaching an advanced age; and because of the vastness of the undertaking the work recedes,' he wrote in 1745; and in 1747, 'My work grows dull... I am overcome by weariness.' The concluding books on feudalism exhaust him. 'This will make three hours' reading; but I assure you that the labor it has cost me has whitened my hair.' 'This work has almost killed me,' he wrote, after revising the final proofs, 'I am going to rest; I shall labour no more.'[59]

In a moving unpublished passage he contemplated the end of his work and his life:

I had conceived the design of giving a much extended and deeper treatment in certain areas of this work; and have become incapable of it. My reading has weakened my eyes, and it seems to me that what remains to me of light is just the dawn of the day when they will close forever.

I am nearly touching the moment when I must begin and end, the moment which unveils and steals everything, the moment mixed with bitterness and joy, the moment when I will lose my very weaknesses.

Why do I still occupy myself with some trivial writings? I search for immortality, and it is within myself. Expand, my soul! Precipitate yourself into immensity! Return to the great Being!...

In the deplorable state in which I find myself it has not been possible for me to give this work its final touches, and I would have burned it a thousand times, if I had not thought it good to render oneself useful to men up to one's very last breath...

Immortal God! the human species is your most worthy work. To love it, is to love you, and, in finishing my life, I devote this love to you.[60]

So what did his mighty labours, the fruits of intense concentration by a first-rate intellect over a period of over 30 years, produce? His mind moved across the data then available for most of the world's civilizations and the whole wealth of human history in order to seek the underlying spirit of the laws, the answer to the riddle of man's nature, past and future.

3
Liberty and Despotism

Montesquieu's central preoccupation was how to maintain liberty and avoid despotism. He was aware from his own experience that liberty was very fragile; Louis XIV had come close to extinguishing it, the Inquisition would do so if it could. Three-quarters of the globe, he thought, suffered from absolutist regimes. Only in Europe had a certain degree of liberty arisen and been preserved. But even here, there was no reason why it should not be extinguished, as in the late history of Rome. His fears are well summarized in the following passage.

> Most of the European nations are still governed by the principles of morality. But if from a long abuse of power or the fury of conquest, despotic sway should prevail to a certain degree, neither morals nor climate would be able to withstand its baleful influence: and then human nature would be exposed, for some time at least, even in this beautiful part of the world, to the insults with which she has been abused in the other three.[1]

The 'long abuse of power' was a recognition of Acton's maxim that 'power tends to corrupt and absolute power corrupts absolutely'.[2] Montesquieu warned that the 'human mind feels such an exquisite pleasure in the exercise of power; even those who are lovers of virtue are so excessively fond of themselves that there is no man so happy as not still to have reason to mistrust his honest intentions.'[3] This was connected to his idea of balance. Something which started as good, balanced and conducive to human happiness and liberty could easily be perverted and swing to a dangerous extreme. For example, if there was no equality between people, democracy was impossible. But if things swung too far the other way, he saw as great a danger. 'The principle of democracy is corrupted not only when the spirit of equality is extinct, but likewise when they fall into a spirit of extreme equality, and when each citizen would fain be upon a level with those whom he has chosen to command him.'[4] Judith Shklar has emphasized Montesquieu's realization of the fragility of liberty. The period of balanced republican liberty, as in Rome, cannot last long. 'The very qualities that make a people prosperous and happy cannot survive in a wealthy and contented society. Nothing seems to fail like republican success.'[5] Thus the 'greatest problem of republic regimes is to put off the evil moment when they lose their inner balance'.[6] Even the freest and most balanced polity he could see, England, would succumb. 'As all human things have an end, the state we are speaking of will lose its liberty... It will perish when all the legislative power shall be more corrupt than the executive.'[7] France was even more in danger. 'What he dreaded was the descent of French absolutism into a despotism on the Spanish model.'[8] Indeed, 'the entire *ancien régime* was at risk... France was structurally inclined towards despotism.'[9] This is a theme also picked up by Shackleton: 'Without vigilance, oriental tyranny might one day govern France. Servitude, says Montesquieu, begins with sleep.'[10] The reason for this is relatively simple. For 'most people are governed by despotism, because any other form of government, any moderate government, necessitates exceedingly careful management and planning, with the most thorough balancing and regulating of political power. Despotism on the other hand, is uniform and simple. Passions alone are required to establish it...'[11]

Montesquieu summarized this inevitable tendency from republic, through monarchy, to despotism, and how the equilibrium cannot be maintained.

> Most European governments are monarchical, or rather are called so; for I do not know whether there ever was a government truly monarchical; at least they cannot have continued very long in their original purity. It is a state in which might is right, and which degenerates always into a despotism or a republic. Authority can never be equally divided between the people and the prince; it is too difficult to maintain an equilibrium; power must diminish on one side while it increases on the other; but the advantage is usually with the prince, as he commands the army.[12]

Control over the army became an increasing threat as technology developed. In particular, Montesquieu noted that the use of gunpowder had further tipped the balance towards despotism: 'You know that since the invention of gunpowder no place is impregnable; that is to say . . . that there is no longer upon the earth a refuge from injustice and violence.'[13] And likewise, 'the invention of bombs alone has deprived all the nations of Europe of freedom' by increasing the need for the centralization of military power.[14]

Liberty could also be lost in 'the fury of conquest'. Montesquieu was aware of two sorts of risk. One was that a country which might have developed internal balance and wealth would be over-run by the 'fury' of conquest by another. He noted how the wealthy civilizations of the Middle East had been 'laid waste by the Tartars, and are still infested by this destructive nation'.[15] Particular danger lay in being part of a continent, not having naturally defensible borders, and being wealthy. His own country of France was a prime example, even when compared to Germany. For 'the Kingdom of Germany was not laid waste and annihilated, as it were, like that of France, by that particular kind of war with which it had been harassed by the Normans and Saracens. There were less riches in Germany, fewer cities to plunder, less extent of coast to scour, more marshes to get over, more forest to penetrate.'[16] Germany's forests and marshes afforded it partial protection, and hence buffered its liberty.

Mountainous regions such as Switzerland had even greater advantages. First, they were poor areas which were not worth attacking, for 'in mountainous districts, as they have but little, they may preserve what they have. The liberty they enjoy, or, in other words, the government they are under, is the only blessing worthy of their defence. It reigns, therefore, more in mountainous and rugged countries than in those which nature seems to have most favoured.'[17] Moderate, non-absolutist, governments were characteristic of mountain areas. 'The mountaineers preserve a more moderate government, because they are not so liable to be conquered. They defend themselves easily, and are attacked with difficulty; ammunition and provisions are collected and carried against them with great expense, for the country furnishes none.'[18]

Even better than mountain barriers was water. Islands were the natural home of liberty. This was not merely because of their defensive advantages. 'The inhabitants of islands have a higher relish for liberty than those of the continent. Islands are commonly of small extent; one part of the people cannot be so easily employed to oppress the other; the sea separates them from great empires; tyranny cannot so well support itself within a small compass: conquerors are stopped by the sea; and the islanders, being without the reach of their arms, more easily preserve their own laws.'[19] This is, implicitly, Montesquieu's major explanation for that puzzle we noted earlier – the different trajectory of England and France. And he explicitly makes the link when showing another advantage of being an island. England, he says, is a nation which 'inhabiting an island, is not fond of conquering, because it would be weakened by distant conquests – especially as the soil of the island is good, for it has then no need of enriching itself by war: and as no citizen is subject to another, each sets a greater value on his own liberty than on the glory of one or any number of citizens'.[20]

The 'fury of war' brings another danger, which Montesquieu showed historically in his account of the way in which incessant aggressive warfare had been at the root of the collapse of liberty in ancient Rome, and also in the ruin caused by Louis XIV's endless wars of attempted conquest. In a section headed 'Of the Augmentation of Troops' Montesquieu described the inevitable Machiavellian law that led continental countries into suicidal wars which then led to higher taxation, poverty and despotism. 'A new distemper has spread itself over Europe, infecting our princes, and inducing them to keep up an exorbitant number of troops. It has its redoublings, and of necessity becomes contagious. For as soon as one prince augments his forces, the rest, of course, do the same; so that nothing is gained thereby but the public ruin.'[21] Thus out of wealth Europe had created poverty and was threatened with the loss of liberty. 'We are poor with the riches and commerce of the whole world; and soon, by thus augmenting our troops, we shall be all soldiers, and be reduced to the very same situation as the Tartars.'[22]

It was a vicious circle: fear–war–higher taxes–absolutism–more fear and so on. 'The consequence of such a situation is the perpetual augmentation of taxes; and the mischief which prevents all future remedy is, that they reckon no more upon their revenues, but in waging war against their whole capital.'[23] Predation was more powerful than production. Indeed, it was a topsy-turvy situation where the richer a country was naturally, the more impoverished and depopulated it would become. 'Most invasions have, therefore, been made in countries which nature seems to have formed for happiness, and as nothing is more nearly allied than desolation and invasion, the best provinces are most frequently depopulated, while the frightful countries of the North continue always inhabited, from their being almost uninhabitable.'[24] According to Montesquieu, after the terrible devastation of the Thirty Years War and then the military adventures of Louis XIV Europe was approaching the situation of India, 'where a multitude of islands and the situation of the land have divided the country into an infinite number of petty states, which from causes that we have not here room to mention are rendered despotic. There are none there but wretches, some pillaging and others pillaged. Their grandees have very moderate fortunes, and those whom they call rich have only a bare subsistence.'[25]

Montesquieu was able to show in detail how the process worked through his study of the rise and decline of the Roman Empire. The essence of the problem was that any success was bound to lead to disaster. It is in the nature of political institutions to grow, and when they do, they lose their way. Thus it was in the very nature of Rome to collapse.

> Rome was made for expansion, and its laws were admirable for this purpose. Thus, whatever its government had been – whether the power of kings, aristocracy, or a popular state – it never ceased undertaking enterprises that made demands on its conduct, and succeeded in them. It did not prove wiser than all the other states on earth for a day, but continually. It sustained meager, moderate and great prosperity with the same superiority, and had neither successes from which it did not profit, nor misfortunes of which it made no use. It lost its liberty because it completed the work it wrought too soon.[26]

He contrasts the situation here with what happens in despotisms. In despotic systems, success makes the despotism ever stronger. In free societies, success inevitably corrupts the freedom.

> What makes free states last a shorter time than others is that both the misfortunes and the successes they encounter almost always cause them to lose their freedom. In a state where the people are held in subjection, however, successes and misfortunes alike confirm their servitude. A wise republic should hazard nothing that exposes it to either good or bad fortune. The only good to which it should aspire is the perpetuation of its condition. If the greatness of the empire ruined the republic, the greatness of the city ruined it no less.[27]

The turning-point in Rome was when she embarked on imperial conquests outside Italy.

> When the domination of Rome was limited to Italy, the republic could easily maintain itself. A soldier was equally a citizen. Every consul raised an army, and other citizens went to war in their turn under his successor. Since the number of troops was not excessive, care was taken to admit into the militia only people who had enough property to have an interest in preserving the city. Finally, the senate was able to observe the conduct of the generals and removed any thought they might have of violating their duty. But when the legions crossed the Alps and the sea, the warriors, who had to be left in the countries they were subjugating for the duration of several campaigns, gradually lost their citizen spirit. And the generals, who disposed of armies and kingdoms, sensed their own strength and could obey no longer.[28]

Montesquieu summarized his findings succinctly. 'Here, in a word, is the history of the Romans. By means of their maxims they conquered all peoples, but when they had succeeded in doing so, their republic could not endure. It was necessary to change the government, and contrary maxims employed by the new government made their greatness collapse.'[29]

The actual process of the fall of Rome, according to Montesquieu, is helpfully summarized by D'Alembert.

> He found the causes of their decadence in the very expansion of the state, which transformed the riots of its people into civil wars; in wars made in places so distant, that citizens were forced into absences of excessive length and lost imperceptibly the spirit essential to republics; in the granting of citizenship to too many nations, and the consequent transformation of the Roman people into a sort of monster with many heads; in the corruption introduced by Asian luxury; in Sulla's proscription, which debased the nation's spirit and prepared it for slavery; in the necessity felt by the Romans, of subjecting themselves to masters, once they felt their liberty to be a burden; in the necessity of changing their maxims along with their form of government; in that series of monsters who reigned almost without interruption from Tiberius to Nerva, and from Commodus to Constantine; and, finally, in the removal and partition of the empire.[30]

What had happened in Rome was the best-documented example of the danger of all continental states. If they were successful, they would have to expand to feed their success and protect their borders. But there were no limits, and as they triumphed on the edges, the centre would become corrupted. It was a phenomenon which Montesquieu had seen in the history of the Spanish Empire and witnessed at first hand under Louis XIV.

Yet there is a contradictory message in Montesquieu as well. Although at times it looks as if the one quarter of the globe that had hints of non-despotic government was tumbling towards the condition of India and China, Montesquieu's argument also rested on the proposition that to a certain extent Europe was still different. He sensed that for the first time in history Europe was becoming the wealthiest and most powerful region in the world, reversing the thousands of years when it had been inferior to the Orient. 'Europe has arrived at so high a degree of power that nothing in history can be compared with it, whether we consider the immensity of its expenses, the grandeur of its engagements, the number of its troops, and the regular payment even of those that are least serviceable, and which are kept only for ostentation.'[31] He thought that this power, and the remnants of the old spirit of liberty, arose from the fact that Europe was divided into a number of roughly equal-sized, equally powerful states, so that no universal despotic Empire could grow up. India faced the problem of political units that were too small; China of units that were too large. Europe, though sucked into incessant wars, at least had some balancing elements.

In a key passage Montesquieu put forward his most favoured theory to account for the difference of Europe and Asia:

> Hence it comes that in Asia the strong nations are opposed to the weak; the warlike, brave, and active people touch immediately upon those who are indolent, effeminate, and timorous; the one must, therefore, conquer, and the other be conquered. In Europe, on the contrary, strong nations are opposed to the strong; and those who join each other have nearly the same courage. This is the grand reason of the weakness of Asia, and of the strength of Europe; of the liberty of Europe, and of the slavery of Asia: a cause that I do not recollect ever to have seen remarked. Hence it proceeds that liberty in Asia never increases; whilst in Europe it is enlarged or diminished, according to particular circumstances.[32]

Asia had been condemned to thousands of years of despotism. Europe vacillated, with liberty rising and collapsing, as Montesquieu had seen both in his studies of Rome and his knowledge of recent European history.

Montesquieu had thus put forward several theories to try to account for the differential success of liberty – geography in particular. Scattered through his works are a number of other theories, at the level both of Europe versus China and of England versus the Continent. One of these lay in the field of religion. At the level of Europe versus Asia, he believed that Christianity in itself was an antidote to despotism. 'The Christian religion is a stranger to mere despotic power. The mildness so frequently recommended in the gospel is incompatible with the despotic rage with which a prince punishes his subjects, and exercises himself in cruelty.'[33]

What he does not seem to have done is to specify, beyond the gospel message of mildness, why Christianity had this effect. Here his experience of the concordat between Church and State in France and most of Catholic Europe may have made him aware that there was nothing intrinsic to Christianity *per se* which would lead it to be a bulwark against state power. He was aware that the mixing of religion and politics was not as extreme as in China. 'The legislators of China went further. They confounded their religion, laws, manners and customs; all these were morality, all these were virtue. The precepts relating to these four points were what they called rites; and it was the exact observance of these that the Chinese Government triumphed.'[34] Yet Christianity could be accommodated into a Caesaro-Papist solution as in Louis XIV's France.

The puzzle was that in one small part of Europe, in the Protestant north-west, there were signs of that desired separation between religion and power. He knew that 'we ought not to regulate by the Principles of the canon Law things which should be regulated by those of the civil Law'.[35] Yet this separation was unusual. Since the thing to be explained was religion, the explanation to the question of why parts of northern Europe was Protestant must lie elsewhere. Montesquieu's favourite explanation seems to have been the climate. His climatic view of religion, which so annoyed the missionaries, applied to religion as a whole. Thus he wrote that 'When a religion adapted to the climate of one country clashes too much with the climate of another it cannot be there established; and whenever it has been introduced it has been afterwards discarded. It seems to all human appearance as if the climate had prescribed the bounds of the Christian and the Mohammedan religions.'[36] Even within Europe, the colder north was more encouraging of liberty, and this liberty and independence led people to want a less centralized and despotic religion. Religion was the consequence of liberty, not the cause. 'The reason is plain: the people of the north have, and will forever have, a spirit of liberty and independence, which the people of the south have not; and, therefore, a religion which has no visible head is more agreeable to the independence of the climate than that which has one.'[37]

This takes us directly on to his climatic arguments for the differential distribution of liberty. Although Montesquieu was not a climatic determinist, he did believe that the different climates both within Europe and as between Europe and Asia explained a good deal. Although he was a southern European from Bordeaux, he had a great deal of respect for northern Europeans, a respect increased during his wide European tour in 1728–31. He came to the conclusion that 'If we travel towards the North, we meet with people who have few vices, many virtues, and a great share of frankness and sincerity. If we draw near the South, we fancy ourselves entirely removed from the verge of morality.'[38] The harsher climate of the north, combined with mountains and poorer soil made the north into a kind of Spartan *dura virum nutrix* (hard nurse of men) which seemed to Montesquieu to lead to hard work and liberty. In India, 'the bad effects of the climate' was 'natural indolence', for the heat led people to want to shun agricultural work.[39] On the other hand, in the north 'The barrenness of the earth renders men industrious, sober, inured to hardship, courageous, and fit for war; they are obliged to procure by labor what the earth refuses to bestow spontaneously.'[40]

That it was not just a harsh climate but the inhospitable and marginal resources that were important is shown by his suggestion that it is on the water-margins of the continents that trade and liberty will flourish, by forcing people into activity. Looking at Europe generally, 'We everywhere see violence and oppression give birth to a commerce founded on economy, while men are constrained to take refuge in marshes, in isles, in the shallows of the sea, and even on rocks themselves. Thus it was that Tyre, Venice, and the cities of Holland were founded.'[41] He therefore began to develop the more general theory that the poorer the resources of a country, the freer the populace. Rich agriculture led to large surpluses which led to predation and hierarchy, either from outsiders or insiders.

'Thus monarchy is more frequently found in fruitful countries, and a republican government in those which are not so; and this is sometimes a sufficient compensation for the inconveniences they suffer by the sterility of the land.'[42] Another way of putting this was to suggest that the 'goodness of the land, in any country, naturally establishes subjection and dependence. The husbandmen, who compose the principal part of the people, are not very jealous of their liberty; they are too busy and too intent on their own private affairs. A country which overflows with wealth is afraid of pillage, afraid of an army.'[43] Thus people who lived on arid mountains, marshes or cold and inhospitable regions would be recompensed by the avoidance of that despotism which lurks on the fertile plains.

As yet this sounds intriguing, but crude and still fairly deterministic. But Montesquieu developed the argument in more complex ways. One was to note the effects of climate and resources on the need for commerce and the division of labour. He noted that there had been a shift in the balance and extent of trade from the south to the north of Europe.

> The ancient commerce, so far as it is known to us, was carried on from one port in the Mediterranean to another; and was almost wholly confined to the South. Now the people of the same climate, having nearly the same things of their own, have not the same need of trading amongst themselves as with those of a different climate. The commerce of Europe was, therefore, formerly less extended than at present.[44]

As an inhabitant of Bordeaux, with its famous medieval wine trade to England, Montesquieu was well aware of the northern need for southern products. However, the 'trade of Europe is, at present, carried on principally from the north to the south; and the difference of climate is the cause that the several nations have great occasion for the merchandise of each other'.[45] The prime example of this development was England. 'As this nation is situated towards the north, and has many superfluous commodities, it must want also a great number of merchandise which its climate will not produce: it has therefore entered into a great and necessary intercourse with the southern nations.'[46] More generally, it was the case that 'In Europe there is a kind of balance between the southern and northern nations. The first have every convenience of life, and few of its wants: the last have many wants, and few conveniences.'[47]

We noted earlier Montesquieu's famous connection between Protestantism (piety), trade (commerce) and liberty. It is worth considering a little further the ways in which he thought commerce was beneficial in relation to liberty. One effect was in reducing war and its destructive effects, 'Peace is the natural effect of trade. Two nations who traffic with each other become reciprocally dependent.'[48] It also encouraged freedom within nations; 'the people of the North have need of liberty, for this can best procure them the means of satisfying all those wants which they have received from nature.'[49] It encouraged freedom from prejudice and good morals. 'Commerce is a cure for the most destructive prejudices; for it is almost a general rule, that wherever we find agreeable manners, there commerce flourishes; and that wherever there is commerce, there we meet with agreeable manners.'[50] It encouraged all the Protestant ethical values, as well as political self-discipline. 'True is it that when a democracy is founded on commerce, private people may acquire vast riches without a corruption of morals. This is because the spirit of commerce is naturally attended with that of frugality, economy, moderation, labor, prudence, tranquillity, order and rule. So long as this spirit subsists, the riches it produces have no bad effect.'[51]

Montesquieu's allusion to the 'corruption of morals' in fact reveals another subtle twist to his argument, for he admitted on other occasions that commerce led to a corruption of morality and in particular to the growth of vanity. Yet even the negative effects could be positive. 'Commercial laws, it may be said, improve manners for the same reason that they destroy them. They corrupt the purest morals.'[52] Fashion or vanity, encouraged by trade, could be extremely beneficial. 'This fashion is a subject of importance; by encouraging a trifling turn of mind, it continually increases the branches of its commerce.'[53] Montesquieu recognized that 'Vanity is as advantageous to a government as pride is dangerous. To be convinced of this we need only represent, on the one hand, the numberless benefits which result from vanity, as industry, the arts, fashions, politeness, and taste.'[54]

In particular, Montesquieu recognized that commerce, leading to manufacture, altered the social hierarchy, leading to a powerful middle class which was a bulwark against tyranny. The contrast between an almost entirely agricultural population such as China or Russia and the powerful middle-class cultures of England or Holland was impressive. 'Commerce itself is inconsistent with the Russian laws. The people are composed only of slaves employed in agriculture, and of slaves called ecclesiastics or gentlemen, who are the lords of those slaves; there is then nobody left for the third estate, which ought to be composed of mechanics and merchants.'[55]

Returning now to the question of the effects of climate and terrain, we can see how complex Montesqieu's reasoning was – for instance we have climate, leading to commerce, encouraging a middle class, which formed a bulwark against despotism. Furthermore, he realized that it was not just the climate itself, but its link to political boundaries that was important. The climatic influence lay as much in the sharp variations of climate within a small area as in the actual climate. The effects would also vary depending on the political boundaries. One of Montesquieu's theories was that if all necessities could be produced within one political boundary, this would lead towards despotism, whereas if necessities had to be exchanged between political entities, this would encourage freedom. Yet it was not just trade in itself. Montesquieu noted the self-sufficiency of Egypt. 'The Egyptians – a people who by their religion and their manners were averse to all communication with strangers – had scarcely at that time any foreign trade. They enjoyed a fruitful soil and great plenty. Their country was the Japan of those times; it possessed everything within itself.'[56] The remark 'The Japan of those times' is intriguing and is expanded elsewhere.

> Let us next consider Japan. The vast quantity of what they receive is the cause of the vast quantity of merchandise they send abroad. Things are thus in as nice an equilibrium as if the importation and exportation were but small. Besides, this kind of exuberance in the state is productive of a thousand advantages; there is a greater consumption, a greater quantity of those things on which the arts are exercised; more men employed, and more numerous means of acquiring power; exigencies may also happen that require a speedy assistance, which so opulent a state can better afford than any other.[57]

This is slightly contradictory, of course, for despite the fact that Montesquieu must have been aware, as he earlier shows, that Japan had in the 1620s closed itself to foreign trade, he talks of 'the vast quantity of merchandise they send abroad'. Yet the passage is valuable as an indication of what Montesquieu thought of the stimulating effects of trade on the arts and on wealth in general.

In another striking passage he describes the connection between liberty and commerce, for not only was trade a cause of liberty, it was also a consequence, for it could only flourish where there was a certain freedom. 'Commerce is sometimes destroyed by conquerors, sometimes cramped by monarchs; it traverses the earth, flies from the places where it is oppressed, and stays where it has liberty to breathe: it reigns at present where nothing was formerly to be seen but deserts, seas, and rocks; and whence it once reigned now there are only deserts.'[58] It was a fragile, fickle, yet powerful force. Just as it could not have its full effect within a large, bounded and inward-turning empire, likewise it was just as harmful if there was too much space between the political entities, as when there was too little. Thus he noted that isolation and poverty were also linked. He believed that the fact that most African coastal societies were still tribal, was principally 'because the small countries capable of being inhabited are separated from each other by large and almost uninhabitable tracts of land'.[59]

There was yet another chain of causation leading from soil and climate towards liberty and this was by way of the agricultural system, in particular the type of crops grown. Montesquieu began to develop a theory that population densities and degrees of liberty were connected. He noted four major agricultural regions in the world and their attendant population densities. 'Pasture-lands are but little peopled, because they find employment only for a few. Corn-lands employ a great many men, and vineyards infinitely more.'[60] This roughly corresponds to the three regions of Europe – the pastoral north, the corn-growing middle and the vineyards of his own Bordeaux and the Mediterranean region, and the increasing population densities associated with them. All of these were compatible with the moderate liberty of republic and monarchy – republics in the pastoral or mixed areas, monarchy in the corn and vineyard areas.

Against the whole of Europe he placed Asia, with the fourth major system, namely wet rice cultivation. He set out the connection between dense population and rice cultivation, particularly in China, in a number of places and it seems to have been one of his implicit explanations for the fact that China seemed to have reached a plateau of wealth and to be the archetype of despotism. He thought that 'China is the place where the customs of the country can never be changed',[61] a place where 'the laws, manners, and customs, even those which seem quite indifferent, such as their mode of dress, are the same to this very day... as they were a thousand years ago.'[62] Since the connections between rice, over-population, and the prevalence of despotism over three-quarters of the globe is central to his argument, it is worth elaborating on his description of the links.

Montesquieu describes the intensive nature of rice cultivation, which both sucks up labour and provides sustenance.

> In countries productive of rice, they are at vast pains in watering the land: a great number of men must therefore be employed. Besides, there is less land required to furnish subsistence for a family than in those which produce other kinds of grain. In fine, the land which is elsewhere employed in raising cattle serves immediately for the subsistence of man; and the labor which in other places is performed by cattle is there performed by men; so that the culture of the soil becomes to man an immense manufacture.[63]

The importance of rice was supplemented by Montesquieu's now rejected views that hot climates led to higher fertility through directly stimulating sexual activity and female reproductive capacity. 'The climate of China is surprisingly favourable to the propagation of the human species. The women are the most prolific in the whole world.'[64] More plausibly, anticipating Smith and Malthus, he argued that there was the universal desire for marriage as soon as possible. 'Wherever a place is found in which two persons can live commodiously, there they enter into marriage. Nature has a sufficient propensity to it, when unrestrained by the difficulty of subsistence.'[65] Rice being so productive and being able to accommodate more and more labour this made early marriage widespread, and hence the population rose rapidly so that 'China grows every day more populous, notwithstanding the exposing of children' and 'the inhabitants are incessantly employed in tilling the lands for their subsistence'.[66]

He added a third reason why the Chinese population grew so quickly, which anticipates more recent theories. This was to do with the attitudes towards the family. Although he does not link this explicitly to Confucianism or the descent system, his observation is perceptive.

> If the population of China is enormous, it is only the result of a certain way of thinking; for since children look upon their parents as gods, reverence them as such in this life, and honour them after death with sacrifices by means of which they believe that their souls, absorbed into Tyen, recommence a new existence,

each one is bent on increasing a family so dutiful in this life, and so necessary for the next.[67]

The negative effect of this was that people lived on the verge of starvation, despite their immense toil and the productiveness of agriculture. 'The people, by the influence of the climate, may grow so numerous, and the means of subsisting may be so uncertain, as to render a universal application to agriculture extremely necessary.'[68] This also suggests that almost all effort had to go into agriculture rather than, as in Europe, into commerce and manufacture. Even thus, 'China, like all other countries that live chiefly upon rice, is subject to frequent famines.'[69] It was a Malthusian treadmill, for 'in China, the women are so prolific, and the human species multiplies so fast, that the lands, though never so much cultivated, are scarcely sufficient to support the inhabitants'.[70] The only compensation was that the blind fury of the populace when famine struck was the one check on the rulers and hence tempered, somewhat, the despotism. 'In spite of tyranny, China by the force of its climate will be ever populous, and triumph over the tyrannical oppressor.'[71] This was because, 'From the very nature of things, a bad administration is here immediately punished. The want of subsistence in so populous a country produces sudden disorders.'[72] Despotism it might be, but a despotism tempered by the need to provide sufficient 'bread and circuses', or rather 'rice and ritual', to stop the millions of long-suffering Chinese from rising to overthrow their masters.

Another argument concerning the relation of physical environment and political absolutism concerns the size of political unit. Montesquieu early developed a sort of political equivalent to the argument that 'small is beautiful'. He summarized his theory succinctly, that 'the natural property of small states to be governed as a republic, of middling ones to be subject to a monarch, and of large empires to be swayed by a despotic prince...'[73] The reasons for this are complex. One seems to be that 'A large empire supposes a despotic authority in the person who governs. It is necessary that the quickness of the prince's resolutions should supply the distance of the places they are sent to.'[74] This link was earlier foreshadowed, as we have seen, in Montesquieu's theory that it was the predatory expansion of Roman civilization outside Italy which inevitably changed it from a republic into a despotic absolutism – a fear which was brought alive again by the aggressive policies of Louis XIV and only avoided by his failures.

The reasons why, in the end, the Roman Empire and despotism collapsed, and why neither the Hapsburgs nor Louis XIV had been able to make Europe into one vast despotic empire, unlike Russia or China, were basically, geographical.

> In Asia they have always had great empires; in Europe these could never subsist. Asia has larger plains; it is cut out into much more extensive divisions by mountains and seas; and as it lies more to the south, its springs are more easily dried up; the mountains are less covered with snow; and the rivers being not so large form more contracted barriers.[75]

For this reason, 'Power in Asia ought, then, to be always despotic: for if their slavery was not severe they would make a division inconsistent with the nature of the country.'[76] In Europe, on the other hand, divided into middle-sized states, rulers have to maintain a balance sufficient to keep the enthusiasm and support of their citizens. 'In Europe the natural division forms many nations of a moderate extent, in which the ruling by laws is not incompatible with the maintenance of the state: on the contrary, it is so favourable to it, that without this the state would fall into decay, and become a prey to its neighbours.'[77] This balance between middle sized political units makes it impossible to set up permanent empires, and encourages liberty. 'It is this which has formed a genius for liberty that renders every part extremely difficult to be subdued and subjected to a foreign power, otherwise than by the laws and the advantage of commerce.'[78]

This analysis has been summarized and commented on by Durkheim, and was important in shaping the latter's thought. He suggests that 'the major role' in Montesquieu's ideas of what shapes the form of a society is played by 'the volume of the society'. In small-scale societies there will be republics, because the 'affairs of the community are at all times present to the eyes and mind of every single citizen'. Thus those in power are only the first among equals.

> But if the society grows larger, everything changes... The increasing differentiation of society gives rise to divergent outlooks and objectives. Further, the sovereign power becomes so great that the person who exercises it is far above all others. The society cannot but change from the republican to the monarchic form.

But if the volume increases still further and becomes excessive, monarchy gives way to despotism, for a vast empire cannot subsist unless the prince has the absolute power enabling him to maintain unity among peoples scattered over so wide an area. So close is the relationship between the nature of a society and its volume that the principle peculiar to each type ceases to operate if the population increases or diminishes excessively.[79]

Durkheim admits that although there are a number of exceptions and objections to this. 'Nevertheless, Montesquieu displays great insight in attributing such influence to the number of social units. This factor is indeed of the highest importance in determining the nature of societies, and in our opinion accounts for the chief differences between them. Religion, ethics, law, the family, etc., cannot be the same in a large society as in a small one.'[80]

For Montesquieu, therefore, there were two levels of analysis. Within the world as a whole, three-quarters was covered by 'despotic' regimes – that is China, India and the Near East, Turkey and Russia. Only western Europe was relatively free, enjoying monarchical and occasionally republican government. This was roughly the same picture that Machiavelli had drawn in the sixteenth century. Although Montesquieu does not explicitly make the link, the line between feudal-monarchical systems, Japan in the East and western Europe in the West, and bureaucratic absolutisms elsewhere was exactly the line of Mongol conquest. Some might suggest a causal connection, lying somewhere in the devastating effect of Mongol invasions on all Montesquieu's middling-level counter-balances to autocratic power, which were removed in each of the Mongol invasions. Although they spared some cities, the Mongols tended to level a great deal, destroying the centre – the towns, universities, small concentrations of wealth and power laboriously built up and left only a massive impoverished peasantry and a vacuum at the top to be filled by an absolute ruler. This happened several times in China, the Middle East and at least once in Russia. But the Mongols stopped on the borders of Austria, and the fleets of Kublai Khan were halted on the beaches of Japan. Only where they did not penetrate did successful commercial capitalist systems with mixed governments develop.

Yet this is moving well beyond Montesquieu – though he hints at this line of argument when he compares the effects of different kind of invasions. In his *Pensées* he contrasted the effect of Islamic and Norse/Germanic invasions, and the innate tendencies towards centralization, working much faster after the former than the latter type of invasion. 'From time to time there take place in the world those inundations of peoples that impose everywhere their customs and mores. The inundation of the Muslims brought despotism; the Northmen, the government of nobles. It took nine hundred years to abolish that government and to establish, in every state, monarchy... That is why there has always been an ebb and flow of empire and liberty.'[81]

Montesquieu drew a second division, within Europe. Anticipating Marc Bloch,[82] he saw that Europe comprised two agrarian civilizations, a Roman-law civilization south of a line running through France and a Germanic common-law civilization in northern France, northern Germany and to the northwards. This was an old and enduring line. There was a further division, between the absolutist-tending monarchies of most of western Europe and the few oases of open, liberal government. It was in two such countries that his books published were published, Holland and in Switzerland. It was in the third, England, that he was able to explore the ways in which peace, prosperity, liberty and piety could all flourish side by side.

4
The Defence of Liberty

Even before he visited England, Montesquieu was aware that it was a country with a peculiarly anti-authoritarian spirit. In the person of Uzbek, he describes the English thus:

> All the nations of Europe are not equally submissive to their princes: the impatient humour of the English, for instance, leaves their king hardly any time to make his authority felt. Submission and obedience are virtues upon which they flatter themselves but little. On this subject they say most amazing things. According to them there is only one tie which can bind men, and that is gratitude: husband and wife, father and son, are only bound to each other by their mutual affection, or by the services they do each other: and these various motives of obligation are the origin of all kingdoms and communities. But if a prince, instead of making the lives of his subjects happy, attempts to oppress and ruin them, the basis of obedience is destroyed; nothing binds them, nothing attaches them to him; and they return to their natural liberty. They maintain that all unlimited power must be unlawful, because it cannot have had a lawful origin. For, we cannot, say they, give to another more power over us than we ourselves have: now, we have not unlimited power over ourselves; for example we have no right to take our own lives: no one upon earth then, they conclude, has such a power.[1]

Referring to the English Civil War, Montesquieu wrote: 'Thus the people of England, finding themselves stronger than one of their kings, pronounced it high treason in a prince to make war upon his subjects.'[2]

It was only after he had visited England, and united his reading of Locke and others with attendance at parliamentary debates and courts of law, that Montesquieu saw how the system worked. His account is famous, and is helpfully summarized by Sorel as follows.

> To make the laws and control their execution, there is a body of legislators composed of representatives of the people elected by a system of suffrage almost universal, for it must include 'all citizens... except those who are in such a low condition that they are considered to have no will of their own;' there is an upper chamber composed of hereditary members sharing with the legislative assembly in making the laws, except those relating to taxes, in regard to which the upper chamber is granted only the right to oppose for fear lest it be corrupted by the crown; there is an executive power entrusted to a monarch, because just as legislation demands deliberation, which is the act of several persons, so execution requires volition, which properly belongs to but one; the executive has not necessarily the power of originating the laws, and takes no part in debates, but has the right to veto new laws; if there is no monarch, the executive power must not be entrusted to members of the legislative assembly, because then the two powers would be blended; the legislative assembly can judge neither the conduct nor the person of the monarch, because this would be a confusion of powers; but though the monarch is inviolable and sacred, his ministers can be called to account and punished. The two chambers meet at stated times, and each year vote on the amount of the taxes and the number of soldiers.[3]

The results of this balance of powers was a set of freedoms which Montesquieu elaborated in several famous chapters of the *Spirit of the Laws*.

There was equality of law. 'Their laws not being made for one individual more than another, each considers himself a monarch; and, indeed the men of this nation are rather confederates than fellow-subjects.'[4] There was religious toleration and liberty.

> With regard to religion, as in this state every subject has a free will, and must consequently be either conducted by the light of his own mind or by the caprice of fancy, it necessarily follows that everyone must either look upon all religion with indifference,

by which means they are led to embrace the established religion, or they must be zealous for religion in general, by which means the number of sects is increased.[5]

There was intellectual freedom. 'As the enjoyment of liberty, and even its support and preservation consist in every man's being allowed to speak his thoughts, and to lay open his sentiments, a citizen in this state will say or write whatever the laws do not expressly forbid to be said or written.'[6] There was freedom to change political allegiance. 'Every individual is independent, and being commonly led by caprice and humour, frequently changes parties.'[7] There was freedom to engage in whatever activity or occupation one liked. In France, for instance, the nobility were kept out of trade. 'In a monarchical government, it is contrary to the spirit of commerce that any of the nobility should be merchants.'[8] The reverse was true in England and this might be a cause as well as a consequence of political freedom. 'It is contrary to the spirit of monarchy to admit the nobility into commerce. The custom of suffering the nobility of England to trade is one of those things which have there mostly contributed to weaken the monarchical government.'[9] In sum, this was a proud and free nation: 'As no subject fears another, the whole nation is proud: for the pride of kings is founded only on their independence. Free nations are haughty; others may more properly be called vain.'[10] All of this meant that it did not really matter how competent the government was. 'In a free nation it is very often a matter of indifference whether individuals reason well or ill; it is sufficient that they do reason: hence springs that liberty which is a security from the effects of these reasonings.'[11]

In many ways Montesquieu's more general definitions of the essence of political liberty were refinements from the actual situation as he perceived it in England. He defined liberty as 'a right of doing whatever the laws permit, and if a citizen could do what they forbid he would no longer be possessed of liberty, because all his fellow-citizens would have the same power'.[12] Liberty and desire should be identical. 'In governments, that is, in societies directed by laws, liberty can consist only in the power of doing what we ought to will, and in not being constrained to do what we ought not to will.'[13] Thus the art of good government was to harmonize individual desire and government policy, using the minimum of force.

> I have often inquired which form of government is most conformable to reason. It seems to me that the most perfect is that which attains its object with the least friction; so that the government which leads men by following their propensities and inclinations is the most perfect. If under a mild government the people are as submissive as under a severe one, the former is to be preferred, since it is more rational, severity being a motive foreign to reason.[14]

There is bound to be some loss of liberty, but it should be minimal. We are told that 'In his notebook he described liberty as a good net in which the fish do not feel constrained.'[15] Political liberty consisted of freedom from fear. 'The political liberty of the subject is a tranquillity of mind arising from the opinion each person has of his safety. In order to have this liberty, it is requisite the government be so constituted as one man need not be afraid of another.'[16] Or again, 'Political liberty consists in security, or, at least, in the opinion that we enjoy security.'[17] 'Philosophic liberty consists in the free exercise of the will; or at least, if we must speak agreeably to all systems, in an opinion that we have the free exercise of our will.'[18]

Thus he believed that 'the safety of the people is the supreme law'[19] and that 'This security is never more dangerously attacked than in public or private accusations. It is, therefore, on the goodness of criminal laws that the liberty of the subject principally depends.'[20] England, with its freedom from torture, freedom from the inquisitorial process of Roman law and the absence of secret accusations was the paramount example of such 'liberty of the subject', so distant from the world of examining magistrates and torture over which Montesquieu in his youth had presided.

The added twist to Montesquieu's argument was that liberty was not merely desirable in itself, but seemed to lead, through trade and manufacture, to economic and thence to political power. Sorel is right that 'Montesquieu did not foresee the speedy advent and prodigious development of modern democracy. Still less would he believe it possible to organize democratic republics in vast countries.'[21] He was also right in detecting a certain disdain in Montesquieu's tone when he wrote 'The politicians of our day talk of nothing but manufactures, commerce, wealth and even luxury!'[22] Yet he is only half right when he continues that 'Montesquieu did not suspect that these manufactures, this commerce, this wealth, and even this luxury, which he considered incompatible with democracies, would one day become their corner-stone, and that this revolution would be effected in his own country and permeate all Europe.'[23]

Montesquieu was living just at the point when it was becoming obvious that England and Holland, and indeed much of north-western Europe, were rapidly becoming both the freest and the richest parts of Europe. This was being achieved by a hitherto untried route. Almost all previous nations had made wealth subservient to power, that is to say predation dominated production. But England had reversed this. 'Other nations have made the interests of commerce yield to those of politics; the English, on the contrary, have ever made their political interests give way to those of commerce.'[24] This is a first hint of the later theme of 'a nation ruled by shopkeepers' which would be fully developed by Adam Smith.

Linked to this reversal was the curious downgrading of the military profession. Unlike every other major western country and particularly the Romans, in England 'Military men are there regarded as belonging to a profession which may be useful but is often dangerous, and as men whose very services are burdensome to the nation: civil qualifications are therefore more esteemed than the military.'[25] Montesquieu noted a similar development in that 'other queen of the sea, the Republic of Holland, so respected in Europe, and so feared in Asia, where its merchants behold many a king bow to the dust before them'.[26]

Montesquieu noted the general shift from the Catholic south to the Protestant north, which he associated with wealth, population and power.

> Before the humiliation of the power of Spain, the Catholics were much stronger than the Protestants. Little by little the latter have arrived at an equality. The Protestants will become richer and more powerful, and the Catholics will grow weaker. The Protestant countries ought to be, and are, in fact, more populous than the Catholic ones; from which it follows, firstly, that their revenue is greater, because it increases in proportion to the number of those who pay taxes; secondly, that their lands are better cultivated; lastly, that commerce is more prosperous, because there are more people who have fortunes to make; and that, with increased wants, there is an increase of resources to supply them.[27]

The last point, concerning the tax base, the conversion of wealth into power, is amplified thus. 'There is no Protestant prince who does not levy upon his people much heavier taxes than the Pope draws from his subjects; yet the latter are poor, while the former live in affluence. Commerce puts life into all ranks among the Protestants, and celibacy lays its hand of death upon all interests among the Catholics.'[28]

The populousness is partly caused, as he notes, by the absence of celibacy. It is also because freedom attracts people. Montesquieu himself had personally experienced this. Not only had he seen the Huguenots flee from France after the Revocation of the Edict of Nantes in 1685, but also in Bordeaux he had seen the after-effects of intolerance of religious minorities in the way the large group of Protestants, including his wife's family, had been treated. Thus he wrote from the heart when he stated,

> The propagation of the species is wonderfully aided by a mild government. All republics are a standing proof of this; especially Switzerland and Holland, which, with regard to the nature of the land, are the two worst countries in Europe, and which are yet the most populous. Nothing attracts strangers more than liberty, and its accompaniment, wealth: the latter is sought after for itself, and our necessity leads us into those countries in which we find the former.[29]

Anticipating much later speculation on the role of marginalized religious minorities such as the Jews or Quakers, he wrote: 'It is worthy of note that those who profess tolerated creeds usually prove more useful to their country than those who profess the established faith; because, being excluded from all honours, and unable to distinguish themselves except by wealth and its shows, they are led to acquire riches by their labour, and to embrace the most toilsome of occupations.'[30] This was another good reason for toleration rather than the oppression he had witnessed in Spain and France.

Montesquieu noticed that England had a very productive agriculture. Such were its surpluses that he relaxed his general rule and admitted that it could afford some luxuries. 'In England the soil produces more grain than is necessary for the maintenance of such as cultivate the land and of those who are employed in the woollen manufactures. This country may be therefore allowed to have some trifling arts and consequently luxury.'[31] The country's obvious wealth was not based on producing unnecessary things, but in particular on its woollen manufacture and trade. 'They enjoy a solid luxury, founded, not on the refinements of vanity, but on that of real wants; they ask nothing of nature but what nature can bestow.'[32] All this was very helpful, and supplemented its greatest natural advantage, which was the fact that it was an island.

> The ruling nation inhabiting a large island, and being in possession of a great trade, has with extraordinary ease grown powerful at sea; and as the preservation of its liberties requires that it should have neither strongholds nor fortresses nor land forces, it has occasion for a formidable navy to defend it against invasions; a navy which must be superior to that of all other powers, who, employing their treasures in wars on land, have not sufficient for those at sea.[33]

This advantage also extended to its overseas policies. It was not, unlike war-weary France, or Italy in the expansion of Rome, lured into the folly of endless land wars. 'This nation, inhabiting an island, is not fond of conquering, because it would be weakened by distant conquests – especially as the soil of the island is so good, for it has then no need of enriching itself by war.'[34] Thus it has become a 'trading people'; 'If this nation sends colonies abroad, it must rather be to extend its commerce than its dominion.'[35]

The final ingredient for England's power arose from a cunning combination of its natural wealth and its freedom. Here Montesquieu developed an interesting idea of the relations between political balance and the use of the citizen's wealth. Montesquieu contrasted this with the usual policy of oppressive governments, as a parable. 'When the savages of Louisiana are desirous of fruit, they cut the tree to the root, and gather the fruit ... This is an emblem of despotic government.'[36] In such a setting of insecurity, as soon as a surplus is generated it is scooped off. This affects everybody, and in particular the chances of developing extensive merchant or manufacturing wealth. 'Hence it is that a merchant under this government is unable to carry on an extensive commerce; he lives from hand to mouth; and were he to encumber himself with a large quantity of merchandise, he would lose more by the exorbitant interest he must give for money than he could possibly get by the goods.'[37]

The high rate of interest was also a result of the insecurity and despotism. 'In those Eastern countries, the greater part of the people are secure in nothing; there is hardly any proportion between the actual possession of a sum and the hopes of receiving it again after having lent it: usury, then, must be raised in proportion to the danger of insolvency.'[38] Ironically, therefore, the tax base shrank and less could be extracted. A vicious circle was entered. Whatever was available was taken away and every sprout was consumed. Hardly the way to encourage manufacture or even agriculture. In such a despotic state 'the incomes of the subjects would cease almost entirely, and consequently that of the prince. There would hardly be any exchange of goods among the citizens, and there would be an end of that circulation of wealth, and of that increase of revenue, which arises from the dependence of the arts upon each other; each person would live upon his land, and would take from it only just enough to keep him from dying of hunger.'[39] He had seen the effects of the consequent downward spiral in the case of Rome. 'No states are in greater need of taxes than those which are growing weaker, so that burdens must be increased in proportion as the ability to pay decreases. Soon, in the Roman provinces, taxes became unbearable.'[40]

The reverse of this was a situation where the citizens were not molested too early and only their surpluses were regularly collected. This was the situation in a place like England or Holland. 'It is a general rule that taxes may be heavier in proportion to the liberty of the subject, and that there is a necessity for reducing them in proportion to the increase of slavery. This has always been and always will be the case.'[41] The regularity and certainty, in other words a fixed amount, was also important – a theme developed by Adam Smith. For 'as the people have a certain knowledge of the necessity of submitting to those taxes, they pay them from the well-founded hope of their discontinuance; their burdens are heavy, but they do not feel their weight: while in other states the uneasiness is infinitely greater than the evil.'[42] Furthermore, since the people identify themselves with their rulers and feel attached to their political system they will, in a free state, make voluntary sacrifices of a kind which are greater than those that can be forced out of them in a despotic one. Again referring to England, he wrote: 'This nation is passionately fond of liberty, because this liberty is real; and it is possible for it, in its defence, to sacrifice its wealth, its ease, its interest, and to support the burden of the heaviest taxes, even such as a despotic prince durst not lay upon his subjects.'[43]

The moral of this was that it was in the interests of both citizens and those in power that wealth should be widely distributed. 'That very equality of the citizens which generally produces equality in their fortunes, brings plenty and vigour into all the parts of the body politic, and spreads these blessings throughout the whole state. It is not so in countries subject to arbitrary power: the prince, the courtiers, and a few private persons, possess all the wealth, while all the rest groan in extreme poverty.'[44] Thus one could conclude 'that if a prince is to be powerful, it is necessary that his subjects should live in luxury; he ought to labour to procure all sorts of superfluities with as much care as the necessities of life.'[45] It was a partnership, rather than an opposition. And from this derived Montesquieu's famous definition of taxation. 'Each citizen contributes to the revenues of the state a portion of his property in order that his tenure of the rest may be more secure.'[46] Furthermore, the citizens trust their government and are therefore prepared to lend it immense sums in its hour of need – which would not be the case in a despotism. Such borrowing unites citizens and State even more closely. 'To preserve its liberty, it borrows of its subjects: and the subjects, seeing that its credit would be lost if ever it were conquered, have a new motive to make fresh efforts in defence of its liberty.'[47]

Thus a free people can attempt tasks apparently well beyond their strength, as Montesquieu had seen with the successes of the Duke of Marlborough in his battles against France, and in the expanding colonies of England. 'It is possible for it to undertake things above its natural strength, and employ against its enemies immense sums of fictitious riches, which the credit and nature of the government may render real.'[48] From very early on the English had enjoyed security of real property from arbitrary seizure by the government 'The Magna Charta of England provides against the seizing of the lands or revenues of a debtor, when his movable or personal goods are sufficient to pay, and he is willing to give them up to his creditors; thus all the goods of an Englishman represented money.'[49] This was one of the reasons behind Montesquieu's affirmation that, as Locke and others had argued, private property should be safeguarded. 'Let us, therefore, lay down a certain maxim, that whenever the public good happens to be the matter in question, it is not for the advantage of the public to deprive an individual of his property, or even to retrench the least part of it by law, or a political regulation.'[50]

*

Montesquieu was thus able to move back and forth between an abstract model of a wealthy, pious and liberal nation and its realization in England. Yet, beyond its island advantage, this still left open the question of why England was so different. Of course, part of the answer lay in geography and climate, but Montesquieu was not prepared to stop there. Reflecting on this very question he wrote, 'I do not deny that the climate may have produced a great part of the laws, manners, and customs of this nation; but I maintain that its manners and customs have a close connection with its laws.'[51] The 'close connection', an anticipation of Weber's 'elective affinity', does not posit a necessary causal chain, but it allows Montesquieu to embark on his clear and sympathetic account of the constitutional and legal arrangements in England, an account which is very heavily based on Locke and Bolingbroke's work.

Montesquieu's greatness, as we noted earlier, lay in his recognition that the solution to many problems lay not in the things themselves but in the relations between things. He was aware that there seemed to be a natural tendency as a nation became wealthier for it to expand and predate on others. This in turn seemed to lead to a growing concentration of power, or, put in another way, a breaking down of the division between spheres, solidifying them into one despotic whole. Freedom and progress, however, consisted in holding them apart.

The normal tendency towards despotism, even within western European nations, is best shown by Montesquieu's study of the Roman Empire. There he reveals how an inevitable pressure occurs. A small nation is successful, but its very success engenders the need to expand farther, and so it goes on. But each expansion shifts the balance away from republic and openness towards a more powerful centre so that one day the people wake up in a totalitarian state. This in turn leads to corruption and final collapse.

The model he developed for China is somewhat different. There was the same move outwards to fill the surrounding vacuum of power until the whole great plain of China was one huge Empire. In this Empire, not only, as we have seen, was religion and polity mixed, but likewise kinship and polity. He noted that 'This empire is formed on the plan of a government of a family. If you diminish the paternal authority, or even if you retrench the ceremonies which express your respect for it, you weaken the reverence due to magistrates, who are considered as fathers...'[52] Thus kinship allegiances strengthen political ties and vice versa. It is a true patriarchal system, power lying in the hands of the father/emperor. The differences between this and the Roman situation is that the Chinese is far more deeply embedded. Partly because of the geography of Europe, partly because of the agriculture, partly because of the differences of religion, China's despotism could not be overturned, unlike Rome's. From time to time China was overrun, by Mongols, Manchus and others. At other times it split into pieces. Yet it always returned quickly to its monolithic shape.

Having explained the interconnectedness of Chinese society and government, Montesquieu continues 'Hence it follows that the laws of China are not destroyed by conquest. Their customs, manners, laws, and religion being the same thing, they cannot change all these at once; and as it will happen that either the conqueror or the conquered must change, in China it has always been the conqueror.'[53] The essence was the absence of a separation of powers, not only between economy, society and polity, but at the level of the rulers, between legislative, executive and judiciary. Thus 'Most kingdoms in Europe enjoy a moderate government' because the prince left the judiciary powers 'to his subjects'. On the other hand, as could be seen, 'In Turkey, where these three powers are united in the Sultan's person, the subjects groan under the most dreadful oppression.'[54]

The secret of liberty was thus firstly a separation of spheres – economy from polity, religion from polity, religion from economy, and society (kinship) from polity, religion and economy. This would be reflected in and re-enforced at the governmental level by the separation and balance between legislature, executive and judiciary. What was needed was both separation and balance. To prevent abuse of power, 'it is necessary from the very nature of things that power should be a check to power. A government may be so constituted, as no man shall be compelled to do things to which the law does not oblige him, nor forced to abstain from things which the law permits.'[55] For instance, 'there is no liberty, if the judiciary power be not separated from the legislative and executive. Were it joined with the legislative, the life and liberty of the subject would be exposed to arbitrary control; for the judge would be then the legislator. Were it joined to the executive power, the judge might behave with violence and oppression.'[56] As Shklar put it, 'The central and continuous theme of the *Spirit of the Laws* is that the independence of the courts of law more than any other institution separates moderate from despotic regimes.'[57] Even within each of these there should be further separations and balances. For instance, as in England with its balance between Commons, Lords and Crown, the 'legislative body being composed of two parts, they check one another by the mutual privilege of rejecting. They are both restrained by the executive power, as the executive is by the legislative.'[58] This idea was partly derived from Locke, but as a number of authors have pointed out Montesquieu went far beyond Locke. He probably owed more to Bolingbroke, and this is a case where a partial myth was created through a creative misreading of the English political system.[59] It would be a fruitful myth, however, for it formed the basis of the American constitution.

The problem, of course, was how to achieve such a balance in the first place and then, more difficult still, how to prevent the balance or dynamic harmony from being lost. Montesquieu had noted the point in Roman history when 'the harmony of the three powers was lost'.[60] Things cannot stand still and the dynamic tension had to be maintained over time. 'These three powers should naturally form a state of repose or inaction. But as there is a necessity for movement in the course of human affairs, they are forced to move, but still in concert.'[61] A middling balance was essential: 'political, like moral good, lying always between two extremes'.[62] Any excess or lurch in one direction, even something intrinsically good, when taken to its extreme, was dangerous. Anticipating Tocqueville, Montesquieu wrote that 'Democracy has, therefore, two excesses to avoid – the spirit of inequality, which leads to aristocracy or monarchy, and the spirit of extreme equality, which leads to despotic power, as the latter is completed by conquest.'[63]

How then could the precarious balance be maintained? Montesquieu suggested that the secret lay in the power of a number of 'intermediary bodies'. We are told that 'The single most important doctrine in *The Spirit of the Laws* is Montesquieu's theory that intermediate bodies like the nobility, the *parlements*, the local courts of seigneurial justice, and the church are all indispensable to political liberty.'[64] As Sorel summarizes Montesquieu's thought,

> It is the nature of monarchy to be founded upon laws. The monarch is the source of all power, political and civil; but he exercises this power by means of channels 'through which his power flows.' These are 'the intermediate, subordinate, and dependent powers,'

moderating 'the shifting and capricious will of a single person.' The two foremost of these powers are the nobility and clergy; the third is a body of magistrates, serving as a repository for constitutional laws, and reminding the prince of them when he seems to forget them. This hierarchy of rank is the necessary condition of monarchical government. If it is destroyed, the inevitable tendency is toward either despotism or democracy.[65]

Thus the nobility administers justice, the *parlements* interpret the laws.[66]

The tension between these is important, for each will be striving for supremacy. But it is essential that none should win completely. Their continued rivalry and inability to become dominant lay behind English liberty. 'The civil power being in the hands of an infinite number of lords, it was an easy matter for the ecclesiastic jurisdiction to gain daily a greater extent. But as the ecclesiastic courts weakened those of the lords, and contributed thereby to give strength to the royal jurisdiction, the latter gradually checked the jurisdiction of the clergy.'[67]

It was like the game of scissors, paper, stone. The clergy checked the lords, the Crown checked the clergy and, presumably, the lords checked the Crown. At a lower level, Montesquieu wrote at length about the need for intermediary bodies, the power of middling level entities such as city corporations, universities, guilds and fraternities. A strong development of such special groups would prevent that despotic division between a single ruler and his court on the one hand, and an enslaved populace on the other, which characterized China and had begun to emerge in Louis XIV's France with the collapse of the provincial and national *parlements* and other intermediary bodies. It was essential that there be many centres of power, each balancing the other.

It was for this reason that Montesquieu argued that, far from showing the imminent collapse of the system, lively confrontations and arguments were a sign of health in a republic. A sustained harmony and peace was the sign of actual or imminent despotism. He showed 'that pluralism and its perpetual tensions and quarrels are the fundamental and necessary conditions of political freedom'.[68] His classic account of this is in relation to Rome. 'We hear in the authors only of the dissensions that ruined Rome, without seeing that these dissensions were necessary to it, that they had always been there and always had to be ... To ask for men in a free state who are bold in war and timid in peace is to wish the impossible. And, as a general rule, whenever we see everyone tranquil in a state that calls itself a republic, we can be sure that liberty does not exist there.'[69]

He then develops the idea of a productive tension, a harmony created through dissonance, a balanced and dynamic equilibrium of forces.

> What is called union in a body politic is a very equivocal thing. The true kind is a union of harmony, whereby all the parts, however opposed they may appear, cooperate for the general good of society – as dissonances in music cooperate in producing overall concord. In a state where we seem to see nothing but commotion there can be union – that is, a harmony resulting in happiness, which alone is true peace. It is as with the parts of the universe, eternally linked together by the action of some and the reaction of others.[70]

By a kind of paradox, the apparent harmony of despotic societies, was actually much more deeply riven by conflict, though the surface was smooth. For, 'in the concord of Asiatic despotism – that is, of all government which is not moderate – there is always real dissension. The worker, the soldier, the lawyer, the magistrate, the noble are joined only inasmuch as some oppress the others without resistance. And, if we see any union there, it is not citizens who are united but dead bodies buried one next to the other.'[71] Balanced polities were vibrant, energetic, noisy, alive; despotism was united by death – a powerful image.

All of this, of course, linked back to his earlier discussions. In theory a virtuous circle was possible. Liberty would encourage trade and manufacture, which would encourage the growth of powerful intermediate power groups, such as those represented in the English House of Commons, which would further establish liberty. Yet Montesquieu was in fact extremely dubious about the sustainability of this circle. The history of the world up to his time showed, as he said, that commercial republics were short-lived. Those of Italy and Germany had collapsed.

> Commercial powers can continue in a state of mediocrity a long time, but their greatness is of short duration. They rise little by little, without anyone noticing, for they engage in no particular

action that resounds and signals their power. But when things have come to the point where people cannot help but see what has happened, everyone seeks to deprive this nation of an advantage it has obtained, so to speak, only by surprise.[72]

It seemed unlikely that Holland or England could long continue to tread the tightrope between success and failure, for every success carried in it the temptation to expand and such expansion would, in the end, lead to collapse.

All that Montesquieu could do was marvel at a current situation where, for a time, a reasonable-sized power seemed to have got the balance right. Thus the English 'which liberty and laws render easy, on being freed from pernicious prejudices, has become a trading people'.[73] And this trading tied in and reinforced the liberty and the religious independence. So that the English 'know better than any other people upon earth how to value, at the same time, these three great advantages – religion, commerce, and liberty'.[74]

*

Montesquieu's studies of the rise and collapse of Roman civilization, and in particular his reading of Caesar and Tacitus on the customs of the Germanic peoples who conquered Rome, suggested an interesting theory to supplement his geographical reasons for the peculiarities of western Europe. In a sense these theories do not replace the earlier arguments. They tend to occur towards the middle and end of *Spirit*, and are an attempt to provide an historical way of looking at the oddness of western Europe.

His basic premise was that in simple tribal societies there were those very values of liberty, equality and fraternity which characterized the best of current nations. Like Rousseau after him, Montesquieu believed that men were, by nature, born free and equal. The simplest people, hunter-gatherers, 'enjoy great liberty; for as they do not cultivate the earth, they are not fixed: they are wanderers and vagabonds; and if a chief should deprive them of their liberty, they would immediately go and seek it under another, or retire into the woods, and there live with their families.'[75] Thus slavery was immoral, for 'as all men are born equal, slavery must be accounted unnatural, though in some countries it be founded on natural reason'.[76] The problem was that what began naturally and could be protected by voting with one's feet, fleeing repression, later had to be protected by artificial means. 'In the state of nature, indeed, all men are born equal, but they cannot continue in this equality. Society makes them lose it, and they recover it only by the protection of the laws.'[77] This, in a nutshell, was the story which he wished to tell in relation to what had happened in western Europe.

Montesquieu's reading of Caesar and Tacitus suggested to him that the early Germanic societies were largely pastoralists, mixing this with hunting and gathering. 'Caesar says, that "The Germans neglected agriculture; that the greatest part of them lived upon milk, cheese, and flesh; that no one had lands or boundaries of his own; that the princes and magistrates of each nation allotted what portion of land they pleased to individuals, and obliged them the year following to remove to some other part."'[78] Or again, 'It seems by Caesar and Tacitus that they applied themselves greatly to a pastoral life; hence the regulations of the codes of barbarian laws almost all relate to their flocks.'[79] Like many pastoral peoples, they were egalitarian and independent-minded, both at the tribal and individual level. They enjoyed a sort of republican structure, a confederation of small chiefdoms with little hierarchy. 'Each tribe apart was free and independent; and when they came to be intermixed, the independency still continued; the country was common, the government peculiar; the territory the same, and the nations different.'[80] Thus they managed to share a territory without becoming locked into an increasingly oppressive state.

They were unusually isolated and rural peoples, as befitted their agriculture, and ruled themselves through a kind of universal suffrage. 'The German nations that conquered the Roman Empire were certainly a free people. Of this we may be convinced only by reading Tacitus "On the Manners of the Germans". The conquerors spread themselves over all the country; living mostly in the fields, and very little in the towns. When they were in Germany, the whole nation was able to assemble.'[81] Any sign of instituted rulers at this time is a mistake. Just as monarchy was absent in much of Europe before the Roman conquests, so 'the peoples of the north and of Germany were not less free; and if traces of kingly government are found among them, it is because the chiefs of armies or republics have been mistaken for monarchs.'[82]

Another odd feature of these early societies was their monetary values in the midst of a pastoral economy. 'The laws of the Germans constituted money a satisfaction for the injuries that were committed, and for the sufferings due to guilt. But as there was but very little specie in the country, they again constituted this money to be paid in goods or chattels.'[83] Curiously for so warlike a peoples, 'Our ancestors, the Germans, admitted of none but pecuniary punishments. Those free and warlike people were of opinion that their blood ought not to be spilled but with sword in hand.'[84] Thus, having very little cash, everything became interchangeable. 'With these people money became cattle, goods, and merchandise, and these again became money.'[85]

Montesquieu's view of the liberating effect of what happened, especially when compared to the effects of the conquests by the Mongols, is summarized as follows.

> Meantime an immense number of unknown races came out of the north, and poured like torrents into the Roman provinces: finding it as easy to conquer as to rob, they dismembered the empire, and founded kingdoms. These peoples were free, and they put such restrictions on the authority of their kings, that they were properly only chiefs or generals. Thus these kingdoms, although founded by force never endured the yoke of the conqueror. When the peoples of Asia, such as the Turks and the Tartars, made conquests, being subject to the will of one person, they thought only of providing him with new subjects, and of establishing by force of arms his reign of might; but the peoples of the north, free in their own countries, having seized the Roman provinces, did not give their chiefs much power. Some of these races, indeed, like the Vandals in Africa and the Goths in Spain, deposed their kings when they ceased to please them; and, amongst others, the power of the prince was limited in a thousand different ways; a great number of lords partook it with him; a war was never undertaken without their consent; the spoils were divided between the chief and the soldiers; and the laws were made in national assemblies. Here you have the fundamental principle of all those states which were formed from the ruins of the Roman Empire.[86]

The first wave of Germanic conquest was later reinforced by a second, with the Vikings. For these Montesquieu has equal praise. He wrote that Scandinavia 'was the source of the liberties of Europe – that is, of almost all the freedom which at present subsists amongst mankind'.[87] Thus, perhaps oddly to us today when we have forgotten that the word French comes from 'Franks', Montesquieu stressed the Germanic roots of not just much of Europe, but in particular France. He wrote that 'Our ancestors, the ancient Germans, lived in a climate where the passions were extremely calm.'[88] He believed that 'it is impossible to gain any insight into our political law unless we are thoroughly acquainted with the laws and manners of the German nations'.[89]

The spread of Germanic civilization helped Montesquieu explain a mystery, that is the uniform and unprecedented spread of an original and new form of civilization in western Europe which grew from the ashes of Roman civilization. 'I should think my work imperfect were I to pass over in silence an event which never again, perhaps, will happen; were I not to speak of those laws which suddenly appeared over all Europe without being connected with any of the former institutions.'[90] In fact, of course, although these laws bore little connection to the Roman civilization which he had studied so closely, they emanated directly from that system described for the Germans by Caesar and Tacitus for 'Such is the origin of the Gothic government amongst us.'[91] This is the system which he admired and whose roots he wished to discover, for they clearly did not lie in Rome. 'The feudal laws form a very beautiful prospect. A venerable old oak raises its lofty head to the skies, the eye sees from afar its spreading leaves; upon drawing nearer, it perceives the trunk but does not discern the root; the ground must be dug up to discover it.'[92]

The discovery of the roots was not merely of antiquarian interest for Montesquieu believed that the quintessence of liberty in modern Europe, that is the separation and balance of powers, had been first expressed in them. And it is therefore not surprising that he should make a great leap across the centuries by joining what he saw in the constitutional balance of early eighteenth century England to what he had read in Tacitus. 'In perusing the admirable treatise of Tacitus 'On the Manners of the Germans', we find it is from that nation the English have borrowed the idea of their political government. This beautiful system was invented first in the woods.'[93] It is not clear from this whether Montesquieu saw a straight continuity, or a conscious reinvention in England. He had neither the sources nor the time to fill in the detail of what happened over the intervening one and a half millenia. This will be a task taken up, as we shall see, by Tocqueville. What is important is to note that Montesquieu's theory gives him not only a stick to beat contemporary absolutisms in Europe with but a hypothesis to explain the differences within Europe.

Drawing on hints in Montesquieu's work, his theory can be put as follows. After the collapse of Rome, much of Europe was covered by a low density Germanic civilization, with its freedom and equality. Then over much of continental Europe, hierarchy and despotism began to reassert itself as a necessary consequence of growing wealth and military confrontation. An expression and re-enforcing of this move towards what Tocqueville would call 'caste' and towards absolutism, was the reintroduction of Roman law and the Roman Catholic religion. In essence Europe lost its freedoms to a resurgent Roman civilization – and this was most evident in southern and central Europe, for instance in France. For reasons which Montesquieu does not elaborate, this returning tide became weaker the further north one went. So England, an island in fact and in law, retained its basically Germanic social structure, political system and monetary values. Thus, with its Germanic Protestantism added to this, it seemed an oasis (with Holland) in a desert of threatened despotism.

Montesquieu had seen the process occur in his studies of Rome; the movement from small, egalitarian, societies, through increasing centralization of power, finally to absolutism and despotism. And he believed he discerned the same process in his own France. He writes of a visit to a library where the histories of all the modern nations are laid out.

> Here are the historians of France, who show us to begin with the power of kings taking shape; then we see it perish twice, and reappear only to languish through many ages; but, insensibly gathering strength and built up on all sides, it achieves its final stage: like those rivers which in their course lose their waters, or hide them under the earth; then reappearing again, swollen by the streams which flow into them, rapidly draw along with them all that opposes their passage.[94]

It culminated in the near-absolutism of Louis XIV when almost all the intermediary, countervailing forces were crushed. In particular, the regional parliaments had withered.

> Parliaments are like those ruins which are trampled under foot, but which always recall the idea of some temple famous on account

of the ancient religion of the people. They hardly interfere now except in matters of law; and their authority will continue to decrease unless some unforeseen event restores them to life and strength. The common fate has overtaken these great bodies; they have yielded to time which destroys everything, to moral corruption which weakens everything, and to absolute power which overbears everything.[95]

What was particularly sad, Montesquieu thought, was that the ancient foundations of freedom in Germanic laws and customs had been lost, and been overlain by the revived, absolutist and imperial, Roman laws. This Roman triumph had been made complete by Roman religion which had joined with Roman law. Speaking of France, Montesquieu asked:

Who would imagine that the most ancient and powerful kingdom in Europe had been governed for ten centuries by laws which were not made for it? If the French had been conquered, it would not be difficult to understand, but they are the conquerors. They have abandoned the old laws made by their first kings in the general assemblies of the nations; and, singularly enough, the Roman laws which have been substituted, were partly made and partly digested by emperors contemporary with their own legislators. And, to make the borrowing complete, and in order that all their wisdom might come from others, they have adopted all the constitutions of the Popes, and have made them a new part of their law: a new kind of slavery.[96]

Montesquieu's historical work was undertaken over two centuries ago. We may wonder how far it stands the test of time, and how far it has been refuted by subsequent research. Here we are fortunate to have a detailed study by Iris Cox on 'Montesquieu and the History of French Laws' which compares his work in great detail with that of more recent scholars. She summarizes her findings thus: 'in my judgement, Montesquieu's historical account stands up well in the light of modern knowledge. His account is comparatively short, but his statements on most of the points he regarded as important in connection with his theory about the spirit of the laws of France are supported in the works to which I have referred.'[97] She lists all his major sections, from the 'organization of early German society, the facts of the Frankish invasion of Gaul' through to 'the gradual re-emergence of Roman law in a different form', and finds that 'all these stages in Montesquieu's outline of development may be found in the pages of Chénon, Lot and other modern historians'.[98] She finds only two matters on which he may be mistaken and which affect his story: 'One is the question as to whether people were free, in Merovingian and Carolingian times, to choose under which law they would live', the other is 'whether, from Merovingian times onwards, the administration of justice was ordinarily attached to the grant of land'.[99] Neither of these possible areas of misinterpretation affect the more general account which I have summarized.

*

When Montesquieu died, his close friend the Earl of Chesterfield wrote the following tribute to him:

> His virtues did honour to human nature; his writings justice. A friend to mankind, he asserted their undoubted and inalienable rights with freedom, even in his own country, whose prejudices in matters of religion and government he had long lamented, and endeavoured, not without some success, to remove. He well knew and justly admired the happy constitution of this country, where fixed and known laws equally restrain monarchy from tyranny, and liberty from licentiousness. His works will illustrate his name and survive him as long as right reason, moral obligation, and the true spirit of laws shall be understood, respected, and maintained.[100]

A century and a half later his French biographer, the historian Sorel, wrote a similar appraisal:

> We have had sublimer philosophers, bolder thinkers, more eloquent writers, sadder, more pathetic, and more fertile creators of fictitious characters, and authors richer in the invention of images. We have had no more judicious observer of human societies, no wiser counsellor regarding great public interests, no man who had united so acute a perception of individual passions with such profound penetration into political institutions, – no one, in short, who has employed such rare literary talent in the service of such perfect good-sense.[101]

These descriptions praise Montesquieu's mixture of high intelligence and courage. He managed to speak out against cruelty, slavery and absolutism despite the dangers. More importantly, he kept his regard for liberty, his freedom of spirit, alive despite the pressures of the French State and the Inquisition. One way to explain his achievement is to take the final message of his *Lettres Persanes*. Throughout the book the absolutist Usbek has been trying to break the spirit of the women in the harem. He believes he has at least achieved this in the case of his favourite, Roxanne, whom he had raped into submission. Then, in the last letter she writes to him as she dies from self-poisoning, she exults that all his oppression has failed. In the midst of tyranny, watched and guarded and punished, she has kept her spirit and soul free. We can hear Montesquieu's voice in hers, as he writes to the absolutist forces of his time. 'How could you think that I was such a weakling as to imagine there was nothing for me in the world but to worship your caprices; that while you indulged all your desires, you should have the right to thwart me in all mine? No: I have lived in slavery, and yet always retained my freedom: I have remodelled your laws upon those of nature; and my mind has always maintained its independence.'[102] Thus it is absolutism which collapses, not the individual conscience. 'For a long time you have had the satisfaction of believing that you had conquered a heart like mine: now we are both delighted: you thought me deceived, and I have deceived you.'[103]

Through his integrity and support for liberty Montesquieu provided a model which would later inspire the two greatest revolutionary movements towards liberty of modern times. He was constantly cited and quoted by the figures in both the American and French revolutions, and Jefferson's declaration of independence and the rights of man was based on his inspiration.

What, then, has this dialogue with Montesquieu contributed to an answer to the two riddles posed at the start of this book, namely what were the forces which have made the development of human civilizations so slow and painful, and then what has enabled the 'progress' of mankind to be so amazingly rapid during the last 300 years? In terms of approach, his elaboration of a comparative, structural and 'ideal type' methodology strengthens the case for believing that these are a fruitful way to approach the problem. He also provides a detailed historical account.

In terms of theory, there are a number of important contributions: the difficulties and precariousness of liberty, the relations of spheres and the tendency towards a rigid overlap of domains, yet the necessity of separation, political pluralism and liberty in order to create a virtuous spiral. Other important ideas include: the dangers associated with being conquered, islands and liberty and the importance of the size of political units on the likelihood of despotism. Above all, Montesquieu lays out a preliminary historical account of how the escape may have happened. He shows the natural tendency of the rise and then destruction of centres of freedom, but he then shows through his examination of European history how England just managed to avoid this tendency. He appreciates the crucial role and nature of feudalism and shows how everywhere (except England) feudalism degenerated into caste and absolutism. He also considers the important case of China.

Among other useful ideas are those concerning the temptations to conquest which ruined the Roman Empire, the importance of Protestant Christianity as a religion of liberty, the importance of climate and the surprising advantages of having poor natural resources, and the advantages of ecological and geographical diversity in encouraging trade. He also noted the beneficial effects of commerce on 'morals', the enormous effects of rice cultivation on population and social structure, the devastating effects of the Mongol invasions, the dangers of over-taxation, and the ways in which liberty in turn, at least for a while, leads to wealth. Finally there is his analysis of the system of checks and balances, of the role of common law and parliament, and of secondary powers, and in the maintenance of liberty once it had been achieved. Ultimately Montesquieu saw the solution to the riddle in the accidental emergence of a balance of powers between lords, clergy, ruler and people, always fragile but somehow long maintained in England.

With Montesquieu we have a first approximation to an answer. Not only do we know how an answer might be constructed, but large parts, particularly on the geographical and historical side, have been partially filled in. Yet a partially completed answer is even more tantalizing and beckons us on to our next encounter with one of Montesquieu's greatest disciples, Adam Smith.

II Wealth

5
Adam Smith's Life and Vision

The historian H.A.L. Fisher summarized the life and influence of Adam Smith as follows:

> A Scot by birth and descent and mixing with the skippers and merchants of Glasgow, where he was long a Professor, he caught the temper of a great seaport struggling against fiscal fetters. His *Wealth of Nations* (1776), the Bible of Economic Science, states in powerful and measured terms the case for Freedom of Trade, and has long governed British policy. Pitt the younger, Huskisson, Peel, Gladstone, Asquith were his pupils. The soul of modesty.[1]

Adam Smith was born in 1723 in Kirkcaldy on the Firth of Forth in Scotland, the son of a Judge Advocate and Comptroller of Customs in the port. His father died just before he was born, and he was brought up by his widowed mother. He went to Kirkcaldy grammar school and then in 1737 to Glasgow University, where he obtained an MA in 1740. In that year he went to Balliol College, Oxford, remaining there for six years. After returning to Scotland he spent two years with his mother and then lived in Edinburgh from 1748 to 1751, where he gave public lectures on literature and jurisprudence. It was at this time that he began what was to become a deep friendship with the philosopher David Hume. Smith was elected professor of logic and then professor of moral philosophy at the University of Glasgow, where he lived from 1751 to 1763. He published his *Theory of Moral Sentiments* in 1759.

In 1764 Smith went to France as tutor of the Duke of Buccleugh and remained there until 1766. He met many of the greatest philosophers and political economists. After a brief stay in London he returned to Kirkcaldy in 1767, and remained there almost without a break until April 1773, working on drafts of *The Wealth of Nations*. He then moved to London and spent a further two and a half years revising the manuscript, which was published in 1776 to great acclaim. From 1778 until his death on 17 July 1790 he lived mainly in Edinburgh, employed as a commissioner of customs and the salt duty. Smith never married; his lifelong companions were his mother, who died in 1787, and his cousin Jane, who died a year later.

It was during the years in Glasgow, as a teacher and also an effective and conscientious administrator, that Smith laid the foundations for his great works. His experience in England had given him a geographical contrast between wealthy England and relatively poor Scotland, but his experience in Kirkcaldy and Glasgow between the 1730s and 1760s gave him an equally important temporal contrast. This was a perfect place from which to witness two dramatic changes. The first was the transformation of the political system. Up to 1745 the older world of the clan system and Catholicism still remained strong in the Highlands as a living contrast to the religious, political and social system of lowland Scotland and England. Then, while Smith was in Oxford, there occurred the last attempt to re-impose this alternative world with the 1745 uprising. When Smith returned in 1746 it was to a country where the clan system was being systematically crushed in the aftermath to Culloden. Smith was thus living on the border of two civilizations and in his formative years watched one of them decisively defeat the other. His unusual insight into the deepest structures of commercial capitalism came out of this experience.

The development of Glasgow itself reinforced this sense of a great shift. Part of the sense of living in two worlds is captured by W.R. Scott when he describes the Glasgow to which Smith returned as professor of moral philosophy. 'The town was his laboratory. In the middle of the eighteenth century it was a remarkable blend of the old and the new. The history of the last hundred years had left enduring traces which were being modified slowly, and sometimes painfully, by a new spirit and by new conditions.'[2] The rapid development of Glasgow in this period is excellently described by Rae, for

> Glasgow had already begun its transition from the small provincial to the great commercial capital, and was therefore at a stage

of development of special value to the philosophical observer. Though still only a quiet but picturesque old place, nestling about the Cathedral and the College and two fine but sleepy streets, in which carriers built their haystacks out before their door, it was carrying on a trade which was even then cosmopolitan. The ships of Glasgow were in all the waters of the world, and its merchants had won the lead in at least one important branch of commerce, the West India tobacco trade, and were founding fresh industries every year with the greatest possible enterprise.[3]

Glasgow was a place where new worlds were being discovered and new methods tried. Smith spent a great deal of time observing and talking to the merchants and made many close friends among them, in particular, Andrew Cochrane, later provost of Glasgow and described by Smollett as 'one of the first sages of the Scottish Kingdom'.[4] Cochrane was clearly a remarkable man;

> Dr. Carlyle tells that 'Dr. Smith acknowledged his obligations to this gentleman's information when he was collecting materials for his *Wealth of Nations* ...' Dr. Carlyle informs us, more-over, that Cochrane founded a weekly club in the 'forties' – a political economy club – of which 'the express design was to inquire into the nature and principles of trade in all its branches, and to communicate knowledge and ideas on that subject to each other,' and that Smith became a member of this club after coming to reside in Glasgow.[5]

Yet Adam Smith would not just have learnt about trade and merchant activities from his Glasgow friends, for there was also rapid industrial development. These new entrepreneurs

> founded the Smithfield ironworks, and imported iron from Russia and Sweden to make hoes and spades for the negroes of Maryland. They founded the Glasgow tannery in 1742, which Pennant thought an amazing sight, and where they employed 300 men making saddles and shoes for the plantations. They opened the Pollokshaws linen printfield in 1742, copper and tin works in 1747, the Delffield pottery in 1748. They began to manufacture carpets and crape in 1759, silk in 1759, and leather gloves in 1763. They opened the first Glasgow bank – the Ship – in 1750, and the second – the Arms – in 1752. They first began

to improve the navigation of the Clyde by the Act of 1759; they built a dry dock at their harbour of Port Glasgow in 1762; while in 1768 they deepened the Clyde up to the city, and began (for this also was mainly their work) the canal to the Forth for their trade with the Baltic. It was obvious, therefore, that this was a period of unique commercial enterprise and expansion.[6]

Thus Adam Smith could see the world changing before his eyes and the affluence he had seen at Oxford spreading rapidly into Scotland. The shock for Smith was made more dramatic in that the very ten years when he had been absent from Glasgow, 1740–50, had witnessed 'a very great change in the appearance of the district. Looking down from the high ground near the University, recently completed mansions of merchants and others in the course of building came into view'.[7] He lived in a boom town and watched a feudal, Calvinist, world dissolving into a commercial capitalist one. The Wealth of Nations is in many ways an almost autobiographical attempt to describe and explain how and why this was happening around him.

He was also living through a kind of experimental test of one of his major theories, namely that free trade and minimal governmental interference would allow the 'natural tendency' for wealth to increase.

When the eighteenth century began, Scotland was excluded, by a series of Acts of the English Legislature, from the Colonial trade. After the Union [in 1723], this restraint was removed, and, by every test, the advance in prosperity, particularly in the West, was remarkable, and even spectacular. Here, then, it seemed that there was something approaching a valid experiment for the verification of an hypothesis, and confirming it up to the hilt.[8]

Smith was living just before the great textile and steam power boom of the later eighteenth century, or the growth of the heavy industries including building of steam ships that would give Glasgow its greatest reputation. This helps to explain certain absences in his work, in particular the omission of the importance of the steam engine. It is important to note this, for another advantage of living in Glasgow and working in the university was that it was the home of several of those who would provide the scientific and technological basis for the industrial revolution. The sort of developments which were happening along the corridor from Smith, and undertaken by friends of his, are described by Rae.

> Only a few years before Smith's arrival they had recognised the new claims of science by establishing a chemical laboratory, in which during Smith's residence the celebrated Dr. Black was working out his discovery of latent heat. They gave a workshop in the College to James Watt in 1756, and made him mathematical instrument maker to the University, when the trade corporations of Glasgow refused to allow him to open a workshop in the city; and it was in that very workshop and at this very period that a Newcomen's engine he repaired set his thoughts revolving till the memorable morning in 1764 when the idea of the separate condenser leapt to his mind as he was strolling past the washhouse on Glasgow Green. They had at the same time in another corner of the College opened a printing office for the better advancement of that art, and were encouraging the University printer, the famous Robert Foulis, to print those Homers and Horaces by which he more than rivalled the Elzevirs and Etiennes of the past.[9]

It is impossible to assess exactly what effect these exciting years in Glasgow had on Smith, but as we shall see, the surviving lecture notes suggest that he had worked out many of his theories to explain the central structural features of a modern commercial economy by 1763. As Stewart observed, 'His long residence in one of the most enlightened mercantile towns in this island, and the habits of intimacy in which he lived with the most respectable of its inhabitants, afforded him an opportunity of deriving what commercial information he stood in need of from the best sources.'[10] And at a deeper level, the contrasts which he could see in time and space, and the sheer rapidity of change, may be behind his great shift to a four-stage model of social development which Meek and others see as one of his greatest contributions. As Meek himself suggests, it seems likely that it was 'the rapidity of contemporary economic advance', and the ease of contrasts between more and less advanced parts of Scotland and England, that gave him the clue. 'If changes in the mode of subsistence were playing such an important and "progressive" role in the development of contemporary society, it seemed a fair bet that they must also have done so in that of past society.'[11]

The situation in Glasgow and around Kirkcaldy was all the more impressive in comparison with the Highlands. As Ross points out 'Britain of the era of the '45 rising provided Smith with a contrast between the Highlands of Scotland at the pastoral stage with a warrior society and patriarchal leaders, and the unwarlike Lowlands, similar to England, organized for agriculture and commerce, and having to rely on a professional army for defence.'[12] Smith developed the theory that discovery and scientific advancement came out of the emotions of 'wonder' and 'surprise', and there was plenty to be amazed at in the Scotland of the 1750s.

Adam Smith had experienced the contrast between a prosperous commercial capitalism in England, and his own poorer Scottish background. This was one precipitant to thought. A second was the contrast within Europe. Early in 1764 having resigned his Professorship, he went to France as tutor of the Duke of Buccleuch, stepson and ward of Charles Townsend. He spent 18 months in Toulouse and then 'About Christmas 1765, they returned to Paris, and remained there till October following. The society in which Mr. Smith spent those ten months, may be conceived from the advantages he enjoyed, in consequence of the recommendations of Mr. Hume.'[13] Smith also visited Geneva, where he made the acquaintance of Voltaire. The correspondence and conversations with Turgot and Quesnai were to be particularly important in his later thought.

The period in Paris is described by Rae.

> Smith went more into society in the few months he resided in Paris that at any other period of his life. He was a regular guest in almost all the famous literary salons of that time ... Our information about his doings is of course meagre, but there is one week in July 1766 in which we happen to have his name mentioned frequently in the course of the correspondence between Hume and his Paris friends regarding a quarrel with Rousseau, and during that week Smith was on the 21st at Mademoiselle l'Espinasse's, on the 25th at Comtesse de Boufflers', and on the 27th at Baron d'Holbach's, where he had some conversation with Turgot. He was a constant visitor at Madame Riccoboni the novelist's. He attended the meetings of the new economist sect in the apartments of Dr. Quesnay, and though the economic dinners of the elder Mirabeau, the "Friend of Men", were not begun for a year after, he no doubt visited the Marquis, as we know he visited other members of the fraternity. He went to Compiegne

when the Court removed to Compiegne, made frequent excursions to interesting places within reach, and is always seen with troops of friends about him.[14]

It is clear that Smith was starting to develop the ideas in his *Lectures on Jurisprudence* into a book when he was in France.

> The Abbé [Morellet] was a metaphysician as well as an economist, but, according to his account of his conversations with Smith, they seem to have discussed mainly economic subjects – 'the theory of commerce,' he says, 'banking, public credit, and various points in the great work which Smith was then meditating,' i.e. the *Wealth of Nations*. This book had therefore by that time taken shape so far that the author made his Paris friends aware of his occupations upon it, and discussed with them definite points in the scheme of doctrine he was unfolding.[15]

Yet it was not just his conversations in Paris that were important. The time in Toulouse may have been equally important; his recent biographer Ross describes how

> It can be argued, however, that for Smith residence in Toulouse yielded an important stock of facts, additional to those collected in Glasgow, about the economic issues that had seized his imagination. These included the division of labour, extent and fluctuation of market, agricultural and commercial systems, the role of transportation in creating wealth, and the struggle for natural liberty in the economic domain. Thus, a walk from the older part of the city to the *quartier parlementaire* to the south provided a lesson in economic history.[16]

After returning to England in 1766, Smith was until early in 1767 an adviser to Charles Townsend, then chancellor of the exchequer, who was preparing his fatal proposal for taxing the American colonies. It is not difficult to see how that experience coloured much of Smith's later writing on taxation. More generally, his spell outside the university was undoubtedly beneficial. The likely effects are summarized by Stewart.

> He had hitherto lived chiefly within the walls of a University, and although to a mind like his, the observation of human nature

on the smallest scale is sufficient to convey a tolerably just conception of what passes on the great theatre of the world, yet it is not to be doubted that the variety of scenes through which he afterwards passed, must have enriched his mind with many new ideas, and corrected many of those misapprehensions of life and manners which the best descriptions of them can scarcely fail to convey.[17]

He returned to Kirkcaldy in May 1767. He had a large stock of data, and a number of theories which he had developed in his lectures. His breadth of reading and personal experience was not limited to Europe. The Americas were of great interest to the Glasgow merchants, but Asia and Africa were also now increasingly attractive. From Montesquieu he had learnt that no theory of how the modern world was developing could omit a consideration of East Asia and particularly China, about which a great deal was being learnt, especially in France. But how to synthesize this immense inflow of rich data and theory?

He thought and wrote for six years and then went to London in April 1773, where he spent another two and a half years revising his manuscript. As Rae states, 'Much of the book as we know it must have been written in London.'[18] He had thought when he arrived in London that the book was almost completed,

> But the researches the author now made in London must have been much more important than he expected, and have occasioned extensive alterations and additions, so that Hume, in congratulating him on the eventual appearance of the work in 1776, wrote, 'It is probably much improved by your last abode in London.' Whole chapters seem to have been put through the forge afresh; and on some of them the author has tool-marked the date of his handiwork himself.[19]

In particular much of the material on America and the colonial experience was added at this stage. As Rae puts it,

> 'We may go further and say that the American Colonies constitute the experimental evidence of the essential truth of the book, without which many of its leading positions had been little more than theory.' It ought of course to be borne in mind that Smith had been in the constant habit of hearing much about the Ameri-

can Colonies and their affairs during his thirteen years in Glasgow from the intelligent merchants and returned planters of that city.[20]

*

In order to understand the changing world around him, Smith refined a number of the theoretical methods which Montesquieu, David Hume and others were advocating. It is arguable that Smith self-consciously set out to apply the Newtonian method to society and history and that he did this with a more than average understanding of what that method was. One part of his debt lay in his use of mechanical analogies in his analysis, which allowed him to investigate the whole of the emerging capitalist and commercial economy and society as if it were some immensely complex mechanism.

Smith used the metaphor of a machine in most of the branches of his analysis. In his early work on the origin of human languages he had likened their structure and progress to that of machines. 'Smith turns to the machine, as he often does seeking explanatory help when describing systems, to provide an analogy for the "progress of language". Original languages have the vast complexity of primitive machines, and both become simpler when gradually the "different parts are more connected and supplied by one another."'[21] Likewise, he applied the idea in relation to the development of scientific or artistic systems or paradigms.

> Systems in many respects resemble machines. A machine is a little system, created to perform, as well as to connect together, in reality, those different movements and effects which the artist has occasion for. A system is an imaginary machine invented to connect together in the fancy those different movements and effects which are already in reality performed.[22]

Or again, society as a whole could be considered as a vast machine. 'Human society, when we contemplate it in a certain abstract and philosophical light, appears like a great, an immense machine, whose regular and harmonious movements produce a thousand agreeable effects.'[23] Finally, the economy could be regarded as a vast machine. He would have been helped in seeing this by the French physiocrats.[24] The analogy gave him confidence that he was investigating an

82 *The Riddle of the Modern World*

infinitely complex set of inter-relations, a structure of some sort. Behind the visible world there lay a set of moving parts, obeying certain rules and principles. This had given Newton his inspiration, and Smith's critics explicitly saw Smith as attempting to discover the 'laws of motion', not of the physical, but of the social and intellectual universe. It was an analytical system which, as Governor Pownall described it, was 'an institute of the Principia *of those laws of motion*, by which the operations of the community are directed and regulated, and by which they should be examined'.[25]

Above all, it replaced God by an invisible hand, the ghost in the machine, to use Koestler's phrase. It incorporated the important idea of the law of unintended consequences into a philosophy which would provide an underpinning for the new world. Meek captures some of this function when he writes that

> the notion that *historical* processes were autonomous but law-governed led to (or was closely associated with) the notion that *economic* processes in a commercial society possessed the same characteristics. The economic 'machine', it was postulated, like the historical 'machine', worked unconsciously but in an orderly and predictable manner to produce results which could be said to be 'subject to law' and which therefore constituted a perfectly proper field of enquiry for the social scientist... The historical machine automatically produced 'progress', which was proclaimed to be (up to a point) a good thing; the economic machine automatically maximised the rate of growth of the national product, which was also proclaimed to be (up to a point) a good thing.[26]

Thus the fact that social, linguistic and economic systems were all like machines, which humans both constructed and improved, made them analysable and guaranteed their 'progress'. And all this happened not through design, but by accident, through unintended consequences. If the world was like a machine, the individuals in it were cogs who, unbeknown to them, were playing an important part in its progress. Thus

> men in pursuing their own objectives seemed frequently to contribute to outcomes which they did not intend or foresee. This doctrine is sometimes described as the law of 'unintended social outcomes' but is more usually cited, in Smith's case, as the doc-

trine of the 'invisible hand' – as in the statement that man is 'led by an invisible hand to promote an end which was no part of his intention'.[27]

Yet even with this confidence that the goal was to work out the laws of motion of an immense machine which lay behind language, thought systems, society and economy, Smith was still faced with the daunting task of devising a method of finding out the hidden principles which drove it. Here again he developed an approach which was in the Newtonian tradition. It appears that Smith first encountered Newton's method in his 'third or *magistrand* year at Glasgow', that is in 1739–40, when he was about 16. The outline of the course then was as follows: '"[the scholars] are taught two Hours at least by the Professor of *Natural Philosophy*, as that science is improved by *Sir Isaac Newton*, and attend two Hours in the Week a Course of Experiments. Some continue to attend Lessons of *Mathematicks*, or of the Lessons of the *Laws of Nature and Nations*, or of *Greek*, or *Latin*." (Chamberlayne, 1737: II.iii.13).'[28] When he returned from Oxford he would have had available the outline of that system published by the recent professor of mathematics at Edinburgh, Colin Maclaurin. In his *Account of Sir Isaac Newton's Philosophical Discoveries* (1748) Maclaurin noted that the scientist

> should begin with phenomena, or effects, and from them investigate the powers or causes that operate in nature; that from particular causes we should proceed to the more general ones, till the argument ends in the most general: this is the method of *analysis*. Being once possessed of these causes, we should then descend in a contrary order, and from them, as established principles, explain all the phenomena that are their consequences, and prove our explications; and this is the *synthesis*. It is evident that, as in mathematics, so in natural philosophy, the investigation of difficult things by the method of *analysis*, ought ever to precede the method of composition, or the synthesis. For in any other way we can never be sure that we assume the principles that really obtain in nature; and that our system, after we have composed it with great labour, is not mere dream and illusion.[29]

In other words there was a backwards and forwards between induction and deduction.

The method required a huge amount of data, for if one were to examine the laws of motion behind the 'great machine' of the world, a wide sweep of materials across time and space were needed, as Montesquieu, whose aims were very similar, had found. As regards space, it was important to consider all kinds of civilization at every level. Thus Smith used his experiences in France, Glasgow and elsewhere, and his collection of travel literature, to learn about China, India, the American Indians and anything else that he could. The work of Du Halde on China and of Lafitau and others on America were especially important, and like Montesquieu he tried to absorb into his general theories the great rush of new knowledge pouring into Europe. He was able to do this the more effectively because, like Montaigne, he maintained a lofty and detached relativism. For example, he recognized that in aesthetics, as well as in everything else, standards were variable and there was nothing that was ultimately 'right'.

> What different ideas are formed in different nations concerning the beauty of the human shape and countenance! A fair complexion is a shocking deformity upon the coast of Guinea. Thick lips and a flat nose are a beauty. In some nations long ears that hang down upon the shoulders are the objects of universal admiration. In China, if a lady's foot is so large as to be fit to walk upon, she is regarded as a monster of ugliness.[30]

He noted that each culture tended to condemn the others as bizarre, without examining their own equally strange practices. Referring to a North American Indian practice of shaping the head into a square form by tying boards round children's heads, he wrote: 'Europeans are astonished at the absurd barbarity of this practice, to which some missionaries have imputed the singular stupidity of those nations among whom it prevails. But when they condemn those savages, they do not reflect that the ladies in Europe had, till within these few years, been endeavouring for near a century past to squeeze the beautiful roundness of their natural shape into a square form of the same kind.'[31]

All this comparative data helped him to fill out the general theory of the evolution of human civilizations which was his life's central work, but even the wealth of material left huge gaps. In particular, it was very difficult to know what had happened in periods before written records survived or in oral cultures. To overcome this problem he developed a method which Dugald Stewart termed 'theoretical or conjectural history'. This was history where, in the absence of direct evidence, as Stewart put it, 'we are under a necessity of supplying the place of fact by conjecture; and when we are unable to ascertain how men have actually conducted themselves upon particular occasions, of considering in what manner they are likely to have proceeded, from the principles of their nature and the circumstances of their external situation.'[32] Campbell and Skinner point out that 'Smith provided early evidence of the technique, opening his treatment of the development of language by supposing two "savages, who had never been taught to speak, but had been bred up remote from the societies of men"'.[33] The method was very close to the Newtonian method. On the basis of the actual evidence one would build up a set of hypotheses or conjectures, moving from the known to the unknown, and then see if these then elicited any further information which refuted or confirmed the 'conjectures'.

*

One of Smith's most famous instances of 'conjectural history' or model building was his elaboration of the four-stage theory of civilization, which has since provided the foundation for all of the social sciences. Basically, he divided the history of civilization into the four 'stages' of hunter-gatherer, pastoralist, settled agriculturalist and 'commercial' society. These stages were defined by the mode of gaining a living and were associated with many other features – the density of population, the development of government, the rise of private property, the development of arts and crafts.[34] Through detailed analysis, Ronald Meek has traced this framework back to Smith's lectures of 1751. This was the very year in which Turgot developed a similar theory, and both of them had been inspired by Montesquieu. Meek believes, however, that Smith took the idea much further than Montesquieu by seeing the stages as naturally developing out of each other, and as primarily determined by the mode of subsistence or earning a living.[35]

Meek is puzzled, however, as to why Smith and Turgot should both seize on a passage in Book XVIII of the *Spirit of the Laws* and simultaneously proceed to transform it dynamically 'into a novel theory of socio-economic development'.[36] It is a puzzle which Meek never really solves, but which we can perhaps resolve by two arguments. First, if we examine Montesquieu more closely, we see that the gap between his theory and Smith's is far less than that assumed by many Smithian scholars. Montesquieu already had a dynamic theory of social change, as in his work on the fall of Rome or the history of feudal Europe. Smith only needed to elaborate a framework that already almost fully existed. Secondly, the feeling of dynamism and 'progress' which we find more pronounced in Smith probably largely reflects his life's experiences. In Montesquieu's home area around Bordeaux, and in France in general, there was only slight 'progress', if any. The great contrast was with England, but that is far less impressive than seeing one's own home area transforming itself within 20 years. As we have seen, this is exactly what Smith observed in the Glasgow region. He felt the massive and rapid shift between stages two and three as the 'pastoral' stage in the Highlands gave way within a couple of decades to the settled agricultural stage with the aftermath of the battle of Culloden in 1745. And he could observe from his windows and talk to the people who were rapidly bringing about a commercial society and laying the groundwork for an industrial one. It is not surprising that his framework should be more 'dynamic' as he watched two of the four great transformations occurring before his eyes.

The importance of this stadial framework was immense. It was the foundation for Smith's thought and that of Ferguson, Millar, Kames and others. It was elaborated and developed by those who refounded the social sciences in the second half of the nineteenth century, strengthened and made into a unified picture of man and nature through the Darwinian vision. It helped provide the framework for the understanding of world history and in particular the mass of new knowledge generated by the expansion of Europe.

*

We see from this account of Smith's life and method that he was well placed to investigate the riddle of the modern world. He lived at its edge, on the border of English civilization, but he also lived at its epicentre, the same corridor where James Watt was revolutionizing the design of the steam engine and hence providing the mechanism to unlock the power to sustain the industrial revolution. Smith was also a part of a great tradition and network of thinkers, from the Greek philosophers onwards.

Although we tend to think of great thinkers as isolated geniuses, of course their views are largely shaped by a complex network of other minds. In many ways, therefore, what we call 'Adam Smith' is a composite, a concentration into one life of numerous trains of thought. Smith synthesized and organized them, but it is no detraction from his greatness to realize that he was just part of a great river through which much of western speculation flowed. Smith undoubtedly owed a heavy debt to the French physiocrats.[37] He owed a good deal to Sir James Steuart.[38] His interests were close to those of Henry Home, Lord Kames; he said that 'we must every one of us acknowledge Kames for our master'.[39] His work overlapped with that of Josiah Tucker.[40] Through his teacher Frances Hutcheson he learnt from the Dutch jurists.

Above all he owed a debt to Hutcheson himself. Five days a week, when he was in his mid-teens studying for his MA in Glasgow, he would attend the inspiring 'prelection' of Hutcheson, covering jurisprudence and politics. Leechman in 1755 described Hutcheson's 'civic humanist theme of the importance of civil and religious liberty for human happiness'.[41]

> as a warm love of liberty, and manly zeal for promoting it, were ruling principles in his own breast, he always insisted upon it at great length, and with the greatest strength of argument and earnestness of persuasion: and he had such success on this important point, that few, if any, of his pupils, whatever contrary prejudices they might bring along with them, ever left him without favourable notions of that side of the question which he espoused and defended.[42]

Smith's work was also connected to that of our other major thinkers, all of whom helped compose the puzzle for him and provided their own answers from which he drew. From Montesquieu's work Smith derived part of his central methodology, in particular the idea of tracing the development of certain institutional forms through a series of stages. Smith was also encouraged to undertake comparative speculation by the *Spirit of the Laws*. A second influence was the poet Alexander Pope, whose *Essay on Man* summarized many of the contradictions between wealth and morality which would lie at the centre of Smith's analysis. The battle between 'self-love' or selfishness and social virtues, constituting the central tension in the emerging capitalist world which Smith would dissect, is Pope's theme, and his resolution of the conflict is almost identical to that proposed by Smith. Pope considers the essence of individualistic capitalism and the way it transmutes selfish competition into public wealth. Both Pope and Smith were influenced by the philosopher Bolingbroke, to whom Pope dedicated the *Essay on Man*. Pope also puts into verse some of the key constituents of Smith's famous theory of the 'invisible hand', that is of the orderliness and fixed laws which lie behind the apparently chaotic flux of events. Many of these themes are also echoed in the work of another writer who influenced Smith, Bernard Mandeville. Mandeville invented the phrase 'the division of labour' and analysed the mechanism in a way that was helpful to Smith. A second overlap was in Mandeville's famous posing of the paradox of 'private vice, public benefit'. Finally, Mandeville argued forcefully for minimal government interference or *laissez-faire*, an idea which Smith explored in greater detail.[43]

The final influence to be considered was the largest of all, namely that of Smith's closest friend, the slightly older philosopher David Hume. They shared many views, for instance on trade, taxation and population growth. One area where Hume tried to resolve a central problem which also lies at the heart of Smith's endeavour was how the production of wealth had grown in Europe despite the strong tendency towards predation on wealth. Hume elegantly outlined the problem, the normal tendency to use any growth in wealth or technology to increase centralized power and social inequality, which in turn would crush further development. Furthermore, he provided an ingenious set of arguments to explain how, once only, the trap had been avoided. He praised manufacture and commerce and explained not only its economic, but also its social and political benefits. He showed how a market economy could become self-sustaining and why international trade benefited everyone. He described the virtuous circle by which increased wealth might lead to a balance of power and to individual liberty, which would in turn lead to further wealth. He showed how a prosperous middle class was central to this process.

Hume's wide-ranging mind also contemplated Montesquieu's question of why Europe seemed to be economically dynamic, while China seemed to be 'stationary', and he put forward several ingenious suggestions which overlap with those which Smith was to develop. He traced the chain of causes from ecology to economy and described the advantages of the political and cultural pluralism of Europe. He showed how such pluralism allowed technological growth and how competition led to wealth accumulation, documenting the advantages of a set of independent units placed within a loosely united civilization. Finally, he explored the question of why England had been more successful than any other country in linking social, religious and political independence to the creation of wealth.[44] Many of these insights were incorporated into Smith's work through a process of discussion and correspondence, and through reading Hume's essays. Yet as with all the thinkers whom Smith encountered, he transmuted and recast their arguments into a new and compelling synthesis which is his own. Given Smith's enormous concentration, deep knowledge and stimulating contacts, we may wonder what his answer was to the riddle of the origins, nature and causes of the central features of a newly emerging world.

6
Growth and Stasis

Adam Smith's economics were based on a broad philosophical tradition concerned with natural law and human nature. The basic premise here, which he derived from Pope, Hutcheson and others, was that man and the natural laws of the universe were in tune. The secret was to release the inhibitions and constraints and then there would be development. Smith's apparent belief in the natural tendency towards progress, and particularly economic progress, is well known. There was the early statement that 'Little else is requisite to carry a State to the highest degree of opulence from the lowest barbarism, but peace, easy taxes, and a tolerable administration of justice; all the rest being brought about by the natural course of things'.[1] Dugald Stewart believed that in his *Wealth of Nations*, Smith had given a 'theoretical delineation' of the 'natural progress of opulence in a country' and the causes, 'which have inverted this order in the different countries of modern Europe'.[2] Furthermore, Stewart elaborated what he saw to be Smith's aim, which was to bring human institutions into line with the 'nature of things', which would then lead to the natural growth of wealth. As Stewart saw it,

> the great and leading object of his speculations is, to illustrate the provision made by nature in the principles of the human mind, and in the circumstances of man's external situation, for a gradual and progressive augmentation in the means of national wealth; and to demonstrate, that the most effectual plan for advancing a people to greatness, is to maintain that order of things which nature has pointed out, by allowing every man, as long as he observes the rules of justice, to pursue his own interest in his own way, and to bring both his industry and his capital into the freest competition with those of his fellow-citizens.[3]

This attempt to free the natural instincts of man lies behind Smith's famous description of man's competitive and rational drives. Smith assumes that the force which leads to the division of labour and accumulation of wealth is 'a certain propensity in human nature ... to truck, barter, and exchange one thing for another'. This is a distinctive and original feature of mankind, connected to the development of reason and speech. 'It is common to all men, and to be found in no other race of animals, which seem to know neither this nor any other species of contracts.'[4] One could go back even further:

> If we should enquire into the principle in the human mind on which this disposition of trucking is founded, it is clearly the natural inclination every one has to persuade. The offering of a shilling, which to us appears to have so plain and simple a meaning, is in reality offering an argument to persuade one to do so and so as it is for his interest. Men always endeavour to persuade others to be of their opinion even when the matter is of no consequence to them.[5]

Once this natural tendency is allowed freedom, the division of labour will mean that 'Every man thus lives by exchanging, or becomes in some measure a merchant, and the society itself grows to be what is properly a commercial society'.[6]

One of Smith's central concerns was to explain why, all else being equal, wealth would grow 'naturally'. One part of his argument lay in a theory of the natural creativity and ingenuity of man, very much along the lines of 'necessity is the mother of invention'. He took it as an axiom that it was a special property of man, as distinct from other animals, to be inventive in relation to technology. 'Man has received from the bounty of nature reason and ingenuity, art, contrivan<c>e, and capacity of improvement far superior to that which she has bestowed on any of the other animalls, but is at the same time in a much more helpless and destitute condition with regard to the support and comfort of his life.'[7] The push towards invention thus came from the fact that humans were so poorly supplied in their natural state.

He pointed to the fact that nearly all inventions came out of a desire to improve the material world. 'Indeed to supply the wants of *meat*, *drink*, cloathing, and lodging allmost the whole of the arts and sciences have been invented and improved.'[8] He asked: 'How many artists are employed to prepare those things with which the shops of the uphorsterrer, the draper, the mercer and cloth-seller <?>, to clip the wool, pick it, sort it, spin, comb, twist, weave, scour, dye, etc. the wool, and a hundred other operators engaged on each different commodities?'[9] This led him to the belief that 'in a certain view of things all the arts, the science<s>, law and government, wisdom, and even virtue itself tend all to this one thing, the providing meat, drink, rayment, and lodging for men, which are commonly reckoned the meanest of employments and fit for the pursuit of none but the lowest and meanest of the people'.[10]

Everything, in the end, came down to practical necessities, to the shifting of atoms. 'Even law and government have these as their finall end and ultimate object. They give the inhabitants of the country liberty and security in the cultivat[ion of] the land which they possess in safety, and their benign influence gives room and opportunity for the improvement of all the various arts and sciences.'[11] He was deeply aware of the complex chain of operations that had led to the apparently simple material objects around him. 'How many have been required to furnish out the coarse linnen shirt [which] he wears; the tanned and dressed-leather-shoes; his bed which he rest<s> in; the grate at which he dresses his victuals; the coals he burns, which have been brought by a long land sea carriage...'[12]

Yet there was still the puzzle of how all these technologies, however desirable, had emerged. Here Smith seems to have made a distinction between small 'micro' inventions, craft or implicit knowledge and 'macro' or 'scientific' or explicit knowledge. He believed that mankind's ingenuity and the concentrated attention which was one consequence of an increasing division of labour would automatically generate small inventions.

> When one is employed constantly on one thing his mind will naturally be employed in devising the most proper means of improving it. It was probably a farmer who first invented the plow, tho the plough wright perhaps, having been accustomed to think on it <?> And there is none of the inventions of that machine so mysterious that one or other of these could not have been the inventor of it. The drill plow, the most ingenious of any, was the invention of a farmer.[13]

He believed that the process was still at work in industrial manufacture. 'But if we go into the work house of any manufacturer in the new works at Sheffiel<d>, Manchester, or Birmingham, or even some towns in Scotland, and enquire concerning the machines, they will tell you that such or such an one was invented by some common workman.'[14]

On the other hand he was also aware that the great 'macro' inventions required a large, non-obvious leap of imagination. They required 'science' or formalized knowledge of some kind. He made the distinction, as often, with an example.

> The wheel wright also, by an effort of thought and after long experien<ce>, might contrive the cog wheel which, turne[n]d by a verticall winch, facilitated the labour exceedingly as it gave the man a superior power over it. But the man who first thought of applying a stream of water and still more the blast of the wind to turn this, by an outer wheel in place of a crank, was neither a millar nor a mill-wright but a philosopher, one of those men who, tho they work at nothing themselves, yet by observing all are enabled by this extended way of thinking to apply things together to produce effects to which they seem noway adapted.[15]

He devoted a good deal of thought to how such 'philosophers' worked and made their discoveries. In essence they had to 'imagine' something that did not yet exist, rather than make minor improvements to what already existed. Their difficulty is summarized as follows, where the process of invention and subsequent refinement is well put.

> The machines that are first invented to perform any particular movement are always the most complex, and succeeding artists generally discover that, with fewer wheels, with fewer principles of motion, than had originally been employed, the same effects may be more easily produced. The first systems, in the same manner, are always the most complex, and a particular connecting chain, or principle, is generally thought necessary to unite every two seemingly disjointed appearances...[16]

Smith believed that humans were naturally curious and ingenious. If these traits were encouraged, there would be rapid technical progress to improve material wellbeing. More difficult to explain were the 'macro' inventions, where he would no doubt have stressed the need for money, a network of contacts, leisure and curiosity.

He also stressed an 'open' intellectual climate, that is a diversity of religious and political opinions, none of them dominant – a competitive world similar to the free trade and competition of the market. 'One thing that has contributed to the increase of this curiosity is that there are now severall sects in Religion and politicall disputes which are greatly dependent on the truth of certain facts.'[17] He had found such a world stimulating in eighteenth-century Scotland.

*

Smith introduced his principle of the division of labour into his lectures sometime in the 1750s. By the time of his lectures on jurisprudence in 1766, he was already using his favourite example, the pin-maker.[18] But he also gave other examples where the division of labour had been in operation. For example, if one took a simple iron tool, 'how many hands has it gone thro. – The miner, the quarrier, the breaker, the smelter, the forger, the maker of the charcoall to smelt it, the smith, etc. have had a hand in the forming it.'[19] This division of labour was the key to improvement and growing opulence: 'It is the great multiplication of the productions of all the different arts, in consequence of the division of labour, which occasions, in a well-governed society, that universal opulence which extends itself to the lowest ranks of the people.'[20] Generalizing from his pin makers, Smith found that

> In every other art and manufacture, the effects of the division of labour are similar to what they are in this very trifling one; though, in many of them, the labour can neither be so much subdivided, nor reduced to so great a simplicity of operation. The division of labour, however, so far as it can be introduced, occasions, in every art, a proportionable increase of the productive powers of labour. The separation of different trades and employments from one another, seems to have taken place, in consequence of this advantage. This separation too is generally carried furthest in those countries which enjoy the highest degree of industry and improvement; what is the work of one man in a rude state of society, being generally that of several in an improved one.[21]

The reasons for the increase in production and hence wealth were threefold.

This great increase of the quantity of work which, in consequence of the division of labour, the same number of people are capable of performing, is owing to three different circumstances; first to the increase of dexterity in every particular workman; secondly, to the saving of the time which is commonly lost in passing from one species of work to another; and lastly, to the invention of a great number of machines which facilitate and abridge labour, and enable one man to do the work of many.[22]

For, as he had put it more succinctly in a lecture twelve years earlier, 'the division of labour increases the work performed from three causes: dexterity acquired by doing one simple thing, the saving of time, and the invention of machines which is occasioned by it.'[23] It is interesting that the third of his reasons was rather different. By breaking up a task into its component parts, it could more clearly be seen where a machine could replace a human being.

The advantages in terms of the improvements in technology, including machinery, were equally important, for the division of labour tended to make micro-inventions more likely. A nice example of how this worked, and to prove his point that most mechanical inventions were made by the workers themselves, was as follows. He described how in the first steam engine,

> a boy was constantly employed to open and shut alternately the communication between the boiler and the cylinder, according as the piston either ascended or descended. One of those boys, who loved to play with his companions, observed that, by tying a string from the handle of the valve which opened this communication to another part of the machine, the valve would open and shut without his assistance, and leave him at liberty to divert himself with his play fellows. One of the greatest improvements that has been made upon this machine, since it was first invented, was in this manner the discovery of a boy who wanted to save his own labour.[24]

Of course, not only is the story, as the footnote to this passage shows, mythical, but Smith omits the fact that in most societies the boy would have put himself out of a job in this way. Yet the more general point concerning the interactions between mechanization and the division of labour is an interesting one. Basically Smith had located the organizational and mechanical side of the industrial revolution. He only lacked the realization of the power of the 'fire engine' as he called it, in other words the steam engine.

He came very close to understanding the potential of new machinery in his early lectures when he discussed the effects of mechanical inventions, and in particular the replacement of human energy by animal, wind and water power.

> The invention of machines vastly increases the quantity of work which is done. This is evident in the most simple operations. A plow with 2 men and three horses will till more ground than twenty men could dig with the spade. A wind or water mill directed by the miller will do more work than 8 men with hand-mills, and this too with great ease, whereas the handmill was reckoned the hardest labour a man could be put to, and therefore none were employed in it but those who had been guilty of some capitall crime. But the handmill was far from being a contemptible machine, and had required a good deal of ingenuity in the invention.[25]

Even with James Watt down the corridor, however, he did not realize the revolution that was just emerging as fossil fuels opened up a vast store of carbon energy.[26]

He did, however, notice a further effect of the division of labour beyond improving production and the chances of mechanical inventions. This was that it led to trade and exchange.

> When the division of labour has been once thoroughly established, it is but a very small part of a man's wants which the produce of his own labour can supply. He supplies the far greater part of them by exchanging that surplus part of the produce of his own labour, which is over and above his own consumption, for such parts of the produce of other men's labour as he has occasion for.[27]

The process was in fact circular and cumulative. Surpluses created exchange or commerce, commerce then encouraged further division of labour and specialization. 'Hence as commerce becomes more and more extensive the division of labour becomes more and more perfect.'[28] He could see this, for instance, within commerce itself if he compared even the fairly advanced parts of Scotland with London: 'A merchant in Glasgow or Aberdeen who deals in linnen will have in his warehouse Irish, Scots, and Hamburgh linnens, but at London there are separate dealers in each of these.'[29] Thus the growth of commercial wealth and division of labour were in many ways two sides of a coin. His dynamic central force was factory production. He had thus almost seen the way in which a commercial economy would break into an industrial one but without the energy revolution could not quite solve the riddle.

*

Smith believed that an increase in commercial activity would lead to an improvement in 'civility'. His prime evidence for its moral effects came from a comparison of various European countries.

> Whenever commerce is introduced into any country, probity and punctuality always accompany it. These virtues in a rude and barbarous country are almost unknown. Of all the nations in Europe, the Dutch, the most commercial, are the most faithfull to their word. The English are more so than the Scotch, but much inferiour to the Dutch, and in the remote parts of this country they <are> far less so than in the commercial parts of it.[30]

He believed that this had nothing to do with race, but rather that self-interest motivated it.

> There is no natural reason why an Englishman or Scotchman should not be as punctual in performing agreements as a Dutchman. It is far more reduceable to self interest, that general principle which regulates the actions of every man, and which leads men to act in a certain manner from views of advantage, and is as deeply implanted in an Englishman as a Dutchman. A dealer is afraid of losing his character, and is scrupulous in observing every engagement.[31]

The reason why the balance shifted from competitive individualism to co-operative behaviour lay in the frequency of transactions. 'Wherever dealings are frequent, a man does not expect to gain so much by any one contract as by probity and punctuality in the whole, and a prudent dealer, who is sensible of his real interest, would rather chuse to lose what he has a right to than give any ground for suspicion.'[32] Thus honesty became the best policy.

Reputation in the longer term was more important than short-term gain.

Smith was also interested in the effects of commerce on art and aesthetics. He started by noting a shift in literary style, from poetry and allusive, dramatic art, to the more practical prose. 'Prose is naturally the Language of Business; as Poetry is of pleasure and amusement. Prose is the Stile in which all the common affairs of Life all Business and Agreements are made.'[33] But he then widened this out to improvements in all arts. 'Tis the Introduction of Commerce or at least of opulence which is commonly the attendent of Commerce which first brings on the improvement of Prose. – Opulence and Commerce commonly precede the improvement of arts, and refinement of every Sort.'[34] He may have had in mind that much art requires leisure, patronage and so on, but there is an implication of something more when he talks of 'refinement'. The mentality bred by commercial societies took them away from the 'rough' manners of warrior societies. He may have had in mind the contrast of the Highland lairds of his youth and contemporary Edinburgh or Glasgow.

Like most great thinkers, Smith's thought arose out of and reflected a series of contradictions. One of these was between his belief that, all else being equal, there was a natural tendency towards the increase of wealth and his realization that in fact such progress only fitfully occurred.[35] This was linked to a second contradiction. Even those countries which seemed to have progressed farthest seemed to have hit some kind of ceiling. In trying to solve these problems he laid the foundations of economics.

Smith did not find it difficult to see why, despite the 'tendency' towards opulence, many societies and civilizations had, in their early stages, remained poor for so long. There was a vicious circle of poverty. 'Bare subsistence is almost all that a savage can procure, and having no stock to begin upon, nothing to maintain him but what is produced by the exertion of his own strength, it is no wonder that he continues long in an indigent state.'[36] Invoking the division of labour, he wrote: 'This is one great cause of the slow progress of opulence in every country; till some stock be produced there can be no division of labour, and before a division of labour take place there can be very little accumulation of stock.'[37] People were forced to share any surplus rather than accumulate a reasonable capital. 'The other arts were all proportionally uncultivated. It was impossible for a man in this state, then, to lay out his whole fortune on himself; the only way . . . to dispose of it was to give it out to others.'³⁸ The ecology was often unimproved and inhospitable; '. . . Tartary and Araby labour under both these difficulties. For in the first place their soil is very poor and such as will hardly admit of culture of any sort, the one on account of its dryness and hardness, the other on account of its steep and uneven surface.'³⁹ Communications were often very poor – particularly with an absence of water transport. 'This is still the case in Asia and other eastern countries; all inland commerce is carried on by great caravans, consisting of several thousands, for mutual defence, with wagons, etca.'⁴⁰

Hovering over all this was constant predation. There was internal predation of the powerful on the weak. 'There could be little accumulation of stock, because the indolent, which would be the greatest number, would live upon the industrious, and spend whatever they produced.'⁴¹ Even if such internal predation could be controlled, there was the danger from foreign invaders. 'Among neighbouring nations in a barbarous state there are perpetual wars, one continualy invading and plundering the other, and tho' private property be secured from the violence of neighburs, it is in danger from hostile invasions. In this manner it is next to impossible that any accumulation of stock can be made.'⁴² He pointed out that 'When people find themselves every moment in danger of being robbed of all they possess, they have no motive to be industrious.'⁴³ He concluded: 'Thus large tracts of country are often laid waste and all the effects carried away: Germany too was in the same condition about the fall of the Roman Empire. Nothing can be more an obstacle to the progress of opulence.'⁴⁴ Nevertheless great civilizations had arisen and overcome these difficulties. Smith pondered about what traps or impediments then lay in their path. Why was the 'tendency' often so weak and ineffective?

Smith's travels on the Continent, his reading of history and accounts of parts of Asia, led him to a number of conclusions. One was that economic development was possible and indeed had occurred in many parts of Europe. One example of this was his own experience in the rapidly expanding economy of lowland Scotland, which was widened when he was able to make an interesting three-way comparison between England, France and Scotland.

> When you go from Scotland to England, the difference which you may remark between the dress and countenance of the common people in the one country and in the other, sufficiently

> indicates the difference in their condition. The contrast is still greater when you return from France. France, though no doubt a richer country than Scotland, seems not to be going forward so fast. It is a common and even a popular opinion in the country, that it is going backwards...[45]

Even though he thought this opinion 'ill founded' he was aware of a check there.

The concept of 'going backwards' is an interesting one, and France is one of his prime examples of a slowing down, if not retreat. While in England the payments for labour had been rising for some time,

> In France, a country not altogether so prosperous, the money price of labour has, since the middle of the last century, been observed to sink gradually with the average money price of corn. Both in the last century and in the present, the day-wages of common labour are there said to have been pretty uniformly about the twentieth part of the average price of the septier of wheat, a measure which contains a little more than four Winchester bushels. In Great Britain the real recompense of labour, it has already been shown, the real quantities of the necessaries and conveniences of life which are given to the labourer, has increased considerably during the course of the present century. The rise in its money price seems to have been the effect, not of any diminution of the value of silver in the general market of Europe, but of a rise in the real price of labour in the particular market of Great Britain, owing to the peculiarly happy circumstances of the country.[46]

Smith was interested in both the cross-sectional wealth of a nation and changes over time – the dynamics of the situation. All of Europe had seen a sudden spurt forward after about 1500, though certain countries had faltered and even 'gone backwards' later. Particularly striking was the shift of gravity from the Mediterranean to the northern countries.

> Since the discovery of America, the greater part of Europe has been much improved. England, Holland, France, and Germany; even Sweden, Denmark, and Russia, have all advanced considerably both in agriculture and in manufactures. Italy seems not to have gone backwards. The fall of Italy preceded the conquest of

Peru. Since that time it seems rather to have recovered a little. Spain and Portugal, indeed, are supposed to have gone backwards. Portugal, however, is but a very small part of Europe, and the declension of Spain is not, perhaps, so great as is commonly imagined.[47]

Thus Italy was more or less stationary, Portugal and Spain declining.

The northern countries had been increasing in wealth but had reached an equilibrium. France had expanded, but by the mid-eighteenth century seemed more or less stationary. Holland, 'in proportion to the extent of the land and the number of its inhabitants', was 'by far the richest country in Europe'.[48] It was 'in proportion to the extent of its territory and the number of its people' a 'richer country than England'. The 'wages of labour are said to be higher in Holland than in England and the Dutch, it is well known, trade upon lower profits than any people in Europe.'[49] Yet Holland also seemed to be stuck, even if it might not be true, as some thought, that it was actually declining.

The one country in Europe which seemed still to be growing rapidly was England.

> Since the time of Henry VIII the wealth and revenue of the country have been continually advancing, and, in the course of their progress, their pace seems rather to have been gradually accelerated than retarded. They seem, not only to have been going on, but to have been going on faster and faster. The wages of labour have been continually increasing during the same period, and in the greater part of the different branches of trade and manufactures the profits of stock have been diminishing.[50]

As Smith surveyed its history all he could see was a gradual but accelerating growth in wealth, century by century, since Roman times.

> The annual produce of the land and labour of England again, was certainly much greater at the restoration [1660], than we can suppose it to have been about an hundred years before, at the accession of Elizabeth [1558]. At this period too, we have all reason to believe, the country was much more advanced in improvement, than it had been about a century before, towards the close of the dissensions between the houses of York and Lancaster. Even then it was, probably, in a better condition than

it had been at the Norman conquest, and at the Norman conquest, than during the confusion of the Saxon Heptarchy. Even at this early period, it was certainly a a more improved country than at the invasion of Julius Caesar, when its inhabitants were nearly in the same state with the savages in North America.[51]

However, Smith's mind ranged beyond Europe. If continued growth was unusual there, how were other areas faring? Here he made a triadic comparison between the New World of North America, the old world of Europe, and the far Eastern world of China. Summing up his impressions of these three, he found that growth was 'rapidly progressive' in North America, 'slow and gradual' in Europe, and 'altogether stationary' in China. The case of the great civilization of China was particularly intriguing and a good negative case to test out his theories. Basing himself on similar sources to those used by Montesquieu, that is the work of Du Halde and the Jesuit missionaries, Smith described the wealthy but stationary state of China, which he thought had roughly existed for at least the last 400 years or so.

> China has been long one of the richest, that is, one of the most fertile, best cultivated, most industrious, and most populous countries in the world. It seems, however, to have been long stationary. Marco Polo, who visited it more than five hundred years ago, describes its cultivation, industry, and populousness, almost in the same terms in which they are described by travellers in the present times. It had perhaps, even long before his time, acquired that full complement of riches which the nature of its laws and institutions permits it to acquire.[52]

Though it was stationary it did not seem to be declining.

> China, however, though it may perhaps stand still, does not seem to go backwards. Its towns are no-where deserted by their inhabitants. The lands which had once been cultivated are no-where neglected. The same or very nearly the same annual labour must therefore continue to be performed, and the funds destined for maintaining it must not, consequently, be sensibly diminished.[53]

Thus it had apparently reached a steady state which others could envy.

Yet while China was 'a much richer country than any part of Europe', not only was it 'stationary' but its common people lived in some hardship.

> The accounts of all travellers, inconsistent in many other respects, agree in the low wages of labour, and in the difficulty which a labourer finds in bringing up a family in China. If by digging the ground a whole day he can get what will purchase a small quantity of rice in the evening, he is contented. The condition of artificers is, if possible, still worse. Instead of waiting indolently in their work-houses, for the calls of their customers, as in Europe, they are continually running about the streets with the tools of their respective trades, offering their service, and as it were begging employment. The poverty of the lower ranks of people in China far surpasses that of the most beggarly nations in Europe.[54]

There were thus two puzzles. One was why had China remained stationary. The other was, why, in such a rich country, were the lower ranks so miserably poor. Smith suggested that the lack of progress was due to the inward-looking, bounded nature of China. The other element was the cultivation of rice, which made labour over-abundant and provided large surpluses which encouraged economic inequality.

The bias towards agriculture and against manufacturing and especially foreign trade was noted by Smith.

> The policy of China favours agriculture more than all other employments. In China, the condition of a labourer is said to be as much superior to that of an artificer; as in most parts of Europe, that of an artificer is to that of a labourer. In China, the great ambition of every man is to get possession of some little bit of land, either in property or in lease; and leases are there said to be granted upon very moderate terms, and to be sufficiently secured to the lessees. The Chinese have little respect for foreign trade. Your beggarly commerce! was the language in which the Mandarins of Peking used to talk to Mr. de Lange, the Russian envoy, concerning it. Except with Japan, the Chinese carry on, themselves, and in their own bottoms, little or no foreign trade; and it is only into one or two ports of their kingdom that they even admit the ships of foreign nations. Foreign trade,

therefore, is, in China, every way confined within a much narrower circle than that to which it would naturally extend itself, if more freedom was allowed to it, either in their own ships, or in those of foreign nations.[55]

It was true that China had a very large internal trade.

> the great extent of the empire of China, the vast multitude of its inhabitants, the variety of climate, and consequently of production in its different provinces, and the easy communication by means of water carriage between the greater part of them, render the home market of that country of so great extent, as to be alone sufficient to support very great manufactures, and to admit of very considerable subdivisions of labour. The home market of China is, perhaps, in extent, not much inferior to the market of all the different countries of Europe put together.[56]

Yet this enormous internal opportunity had turned the Chinese inwards. Without foreign trade they became too bounded and unable to benefit from external ideas and improvements. The following passage, which echoes to a considerable extent the ideas of Du Halde upon whom Smith was dependent, summarizes one of Smith's main theories to account for China's stagnation, and by implication one of the reasons for Europe's relative dynamism.

> A more extensive foreign trade, however, which to this great home market added the foreign market of all the rest of the world; especially if any considerable part of this trade was carried on in Chinese ships; could scarce fail to increase very much the manufactures of China, and to improve very much the productive powers of its manufacturing industry. By a more extensive navigation, the Chinese would naturally learn the art of using and constructing themselves all the different machines made use of in other countries, as well as the other improvements of art and industry which are practised in all the different parts of the world. Upon their present plan they have little opportunity of improving themselves by the example of any other nation; except that of the Japanese.[57]

Another thing Smith noted, like Montesquieu, was that the fruitfulness of rice led to a very dense population. 'In rice countries, which generally yield two, sometimes three crops in the year, each of them more plentiful than any common crop of corn, the abundance of food must be much greater than in any corn country of equal extent. Such countries are accordingly much more populous.'[58] One consequence was that the rich could purchase large numbers of followers in a way that was impossible in Europe. Wealthy people

> having a greater super-abundance of food to dispose of beyond what they themselves can consume, have the means of purchasing a much greater quantity of the labour of other people. The retinue of a grandee in China or Indostan accordingly is, by all accounts, much more numerous and splendid than that of the richest subjects in Europe.[59]

Smith also notes that China and India though not 'much inferior' were definitely 'inferior' in their 'manufacturing art and industry' to Europe.[60] At this point he does not make any connection between the availability of very cheap and plentiful labour and the relative inferiority of manufacturing and machinery.

A second consequence of rice cultivation was that it encouraged extreme social stratification, a class of landlords. This was again because of the bountifulness of rice.

> Though its cultivation, therefore, requires more labour, a much greater surplus remains after maintaining all that labour. In those rice countries, therefore, where rice is the common and favourite vegetable food of the people, and where the cultivators are chiefly maintained with it, a greater share of this greater surplus should belong to the landlord than in corn countries.[61]

The other effect of the bountifulness of rice is to produce not only a very dense population, but one which will continue to grow ever more dense at every opportunity. In this way, as Smith noted, it tends to have the same effect as potatoes. Because of the much higher food value of potatoes, 'much superior to what is produced by a field of wheat', Smith thought that

> Should this root ever become in any part of Europe, like rice in some rice countries, the common and favourite vegetable food of the people, so as to occupy the same proportion of the lands in tillage which wheat and other sorts of grain for human food

do at present, the same quantity of cultivated land would maintain a much greater number of people, and the labourers being generally fed with potatoes, a greater surplus would remain after replacing all the stock and maintaining all the labour employed in cultivation. A greater share of this surplus too would belong to the landlord. Population would increase, and rents would rise much beyond what they are at present.[62]

Thus one would simultaneously have richer landlords and a swarming population. The same was true of rice. It encouraged the population to rise. 'Marriage is encouraged in China.'[63] If he had been able to obtain better data he would have realized that since the late seventeenth century Chinese population had risen very fast. What he was roughly describing in his description of a long stationary period is the famous 'high-level equilibrium trap'.[64]

7
Of Wealth and Liberty

One of the conditions for growth was the development of towns. Adam Smith's experience in Glasgow, where he could see before his eyes the effect of the rapid growth of a city and could talk to prosperous manufacturers and traders, gives his account of the role of towns in economic growth a particular depth and interest. It is also fascinating because it is so deeply ambivalent and contradictory, both laudatory and condemning of this growth. In a chapter significantly entitled 'Of the Natural Progress of Opulence', he started by pointing out that towns were important to commercial development.

> The great commerce of every civilized society, is that carried on between the inhabitants of the town and those of the country. It consists in the exchange of rude for manufactured produce, either immediately, or by the intervention of money, or of some sort of paper which represents money. The country supplies the town with the means of subsistence, and the materials of manufacture. The town repays this supply by sending back a part of the manufactured produce to the inhabitants of the country.[1]

This was welcome for 'The gains of both are mutual and reciprocal, and the division of labour is in this, as in all other cases, advantageous to all the different persons . . .'[2] Elsewhere he pointed out, in a chapter titled 'How the Commerce of the Towns Contributed to the Improvement of the Country', that there were three effects on the countryside. As he put it in the marginal headings, these were 'because they afforded (1) a ready market for its produce (2) because merchants bought land in the country and improved it and (3) because order and good government were introduced'.[3]

It is worth quoting Smith a little further. He wrote that

> commerce and manufactures gradually introduced order and good government, and with them, the liberty and security of individuals, among the inhabitants of the country, who had before lived almost in a continual state of war with their neighbours, and of servile dependency upon their superiors. This, though it has been the least observed, is by far the most important of all their effects. Mr. Hume is the only writer who, so far as I know, has hitherto taken notice of it.[4]

The bands of retainers were dismissed and the lords became prosperous capitalists.

Smith was fully aware that free trading and manufacturing towns were unlikely to emerge from agrarian civilizations. Foreshadowing Marx and Weber he gives an excellent sketch of their chance emergence and their peculiarity in the West. He describes how after the Fall of the Roman Empire 'Free Burghs' began to emerge in the West, having control over their own taxation and their own government.

> They were gradually at the same time erected into a commonality or corporation, with the privilege of having magistrates and a town-council of their own, of making bye-laws for their own government, of building walls for their own defence, and of reducing all their inhabitants under a sort of military discipline, by obliging them to watch and ward; that is, as anciently understood, to guard and defend those walls against all attacks and surprises by night as well as by day. In England they were generally exempted from suit to the hundred and country courts; and all such pleas as should arise among them, the pleas of the crown excepted, were left to the decision of their own magistrates. In other countries much greater and more extensive jurisdictions were frequently granted to them.[5]

Such a development was amazing. For instance, in relation to their ability to tax themselves, it was extraordinary that the sovereigns of all the different countries of Europe

> should have exchanged in this manner for a rent certain, never more to be augmented, that branch of the revenue, which was, perhaps, of all others the most likely to be improved by the

natural course of things, without either expense or attention of their own: and that they should, besides, have in this manner voluntarily erected a sort of independent republics in the heart of their own dominions.[6]

Given the possibility of predating on this, why were they set free?
Here Smith develops the ingenious theory that basically they managed to escape through the tension between the King and his feudal nobles. His account of this process, whereby the King sided with the towns in his battles with the nobles, is worth giving in full. Starting with the feudal lords, he noted that

> the wealth of the burghers never failed to provoke their envy and indignation, and they plundered them upon every occasion without mercy or remorse. The burghers naturally hated and feared the lords. The king hated and feared them too; but though perhaps he might despise, he had no reason either to hate or fear the burghers. Mutual interest, therefore, disposed them to support the king, and the king to support them against the lords. They were the enemies of his enemies, and it was his interest to render them as secure and independent of those enemies as he could. By granting them magistrates of their own, the privilege of making bye-laws for their own government, that of building walls for their own defence, and that of reducing all their inhabitants under a sort of military discipline, he gave them all the means of security and independency of the barons which it was in his power to bestow.[7]

The support of the King built up the strength of those who lived in the towns, forming them into a separate and powerful estate of their own.

> These burghers were such, and were therefore greatly encouraged by them, and we find accordingly that all the burghers and freed sort of slaves who lived in the villages or towns, which any villain became who left his master and lived in one of these towns for a year without being claim'd, had the liberty of marrying whom they pleased, of free trade, etc., without any toll. They were afterwards formed into corporations holding in capite [directly] of the king, having a jurisdiction and territory for which they paid a certain rent.[8]

As their power grew, they began to defend themselves against the predations of local lords.

> In this manner these small towns became free and able to protect themselves, as they had a stout stone wall about the town and kept a constant watch and ward, which was one part of the duty of a burgher, and were always ready for arms and battle to defend themselves against the attempts of the lords, who frequently disturbed them and often plundered their towns.[9]

The danger, however, was that they would go too far in their independence. If they lost their alliance with the ruler, they might prosper for a time. This was exactly what happened in Italy and Switzerland, where 'on account either of their distance from the principal seat of government, of the natural strength of the country itself, or of some other reason, the sovereign came to lose the whole of his authority, the cities generally became independent republics, and conquered all the nobility in their neighbourhood; obliging them to pull down their castles in the country, and to live, like other peaceable inhabitants, in the city'.[10] But in the long run they were too small to be viable and were finally crushed by foreign invaders, as in Italy.

In France and England, however, 'the cities had no opportunity of becoming entirely independent'. Yet they jealously preserved some autonomy and, for instance, 'the sovereign could impose no tax upon them...without their own consent'.[11] Thus they emerged as expanding oases of order and rational wealth production in an agrarian landscape otherwise characterized by predatory, feuding lords. As Smith put it in the marginal heading: 'In consequence of this greater security of the towns industry flourished and stock accumulated there earlier than in the country.' Thus 'Order and good government, and along with them the liberty and security of individuals, were, in this manner, established in cities, at a time when the occupiers of land in the country were exposed to every sort of violence.'[12] Art and good manners also flourished. 'Wherever the Inhabitants of a city are rich and opulent, where they enjoy the necessaries and conveniences of life in ease and Security, there the arts will be cultivated and refinement of manners a neverfailing attendent.'[13] Thus, as many later commentators would argue, the growth of towns and the growth of commercial capitalism went hand in hand, and Smith has given some hints why in the fragmented and balanced politics of Europe a type of 'free town' should emerge which later Weber was to show was entirely different to that in the absolutist empires of the east.[14]

So what was Smith's objection to towns? He thought, ultimately, that town and countryside would become opposed. He believed, as he put it, that agriculture was primary and trade and town manufacture was secondary.

> As subsistence is, in the nature of things, prior to convenience and luxury, so the industry which procures the former, must necessarily be prior to that which ministers to the latter. The cultivation and improvement of the country, therefore, which affords subsistence, must, necessarily, be prior to the increase of the town, which furnishes only the means of convenience and luxury.[15]

He also believed that human beings preferred living in the country and would move there if they made sufficient money in the towns. Thus the 'natural order of things', was for the countryside to flourish, and then the towns to follow suit. 'Had human institutions, therefore, never disturbed the natural course of things, the progressive wealth and increase of towns would, in every political society, be consequential, and in proportion to the improvement and cultivation of the territory or country.'[16]

Another concern was that, though entrepreneurs ought to prefer the security of manufacturing goods for use in their own country rather than in getting involved in highly risky foreign trade, and should prefer agriculture to trade, yet this 'normal' course was increasingly being perverted in eighteenth-century Europe. For 'though this natural order of things must have taken place in some degree in every such society, it has, in all the modern states of Europe, been, in many respects, entirely inverted. The foreign commerce of some of their cities has introduced all their finer manufactures, or such as were fit for distant sale; and manufactures and foreign commerce together, have given birth to the principle improvements of agriculture.'[17] In fact Glasgow was a prime example of the reversal of this 'natural' order, and hence to be castigated, since its wealth was principally based, like that of Holland, on long-distance trade – in particular, as we have seen, the tobacco and other trades with the West Indies and the southern states of America, and on slavery.

This was closely linked to Smith's ambivalent attitude to merchants and manufacturers. On the one hand they were the focus for the first development of commercial capitalism, of liberty, and of the subduing of violence through the spread of wealth, and as such they are the heroes of his story. On the other hand he had no illusions about their benevolence. They had emerged by complete chance out of the stand-off between feudal lords and kings. As Dugald Stewart summarized Smith's position, the emergence of commercial centres 'took their rise, not from any general scheme of policy, but from the private interests and prejudices of particular orders of men'. This 'state of society, however, which at first arose from a singular combination of accidents, has been prolonged much beyond its natural period, by a false system of Political Economy, propagated by merchants and manufacturers, a class of individuals whose interest is not always the same with that of the public...' In other words, they had become too powerful – oligarchic and monopolistic and guild bound. Thus 'By means of this system, a new set of obstacles to the progress of national prosperity has been created.'[18]

In particular, Smith was alluding to trade restrictions based on his enemy, the mercantilist philosophy. 'The false system of Political Economy which has been hitherto prevalent, as its professed object has been to regulate the commercial intercourse between different nations, has produced its effect in a way less direct and less manifest, but equally prejudicial to the states that have adopted it.'[19] Thus the uneven development of the three components – agriculture, manufacture and trade, had led to the development of what Smith called the 'Commercial' or 'Mercantile' System or what we call Mercantilism. The two main methods of enriching a nation under this system were

> restraints upon importation, and encouragements to exportation. Part of these expedients, he observes, have been dictated by the spirit of monopoly, and part by a spirit of jealousy against those countries with which the balance of trade is supposed to be disadvantageous. All of them appear clearly, from his reasonings, to have a tendency unfavourable to the wealth of the nation which imposes them.[20]

Thus Smith felt that a good deal of the independent power of towns and their inhabitants was a beneficial accident in the west, but that the development was going too far towards trade monopolies and sectional interests.

In considering the problem of why England's wealth had 'insensibly' crept up and continued to grow, one key, Smith believed, lay in the social structure. His model of the economy and society is extremely 'modern'; it is not based on the usual *Ancien Régime* structure of a number of legally separate 'estates' of nobility, peasantry, clergy and bourgeois, who exchange goods and services. It is split into 'three different orders of people ... those who live by rent ... by wages ... by profit. These are the three great, original and constituent orders of every civilized society'.[21] They are the landlords, wage-labourers and employers of our modern capitalist state. It is clear from his analysis that he built this model up on the basis of his observations of how English society worked.

When trying to explain why England was so successful, he considered its geographical advantages, agreeing that it is 'perhaps as well fitted by nature as any large country in Europe, to be the seat of foreign commerce'.[22] He also pointed out that its legal code was favourable to commerce: 'in reality there is no country in Europe, Holland itself not excepted, of which the law is, upon the whole, more favourable to this sort of industry...'[23] But the geographical and legal advantages were less important than one other; 'what is of much more importance than all of them, the yeomanry of England are rendered as secure, as independent, and as respectable as law can make them.'[24] In other words, it is the curious position of what roughly might be called 'the middle class' that is crucial.

Smith asks rhetorically what would the position of England have been if it 'had left the yeomanry in the same condition as in most other countries of Europe?'[25] He believed that 'Those laws and customs so favourable to the yeomanry, have perhaps contributed more to the present grandeur of England, than all their boasted regulations of commerce taken together.'[26] For their position and status was very different in England. 'Through the greater part of Europe the yeomanry are regarded as an inferior rank of people, even to the better sort of tradesmen and mechanics...'[27] There is consequently little investment by townsmen in the countryside, he believed, except in England, Holland and Berne in Switzerland.

As to why the yeomanry should be so powerful and prosperous, Smith's answer seems to be that in England, above all, the property law was such that they had private property and security of tenure. Even leases are more secure than elsewhere.

In England, therefore, the security of the tenant is equal to that of the proprietor. In England besides a lease for life of forty shillings a year value is a freehold, and entitles the lessee to vote for a member of parliament; and as a great part of the yeomanry have freeholds of this kind, the whole order becomes respectable to the landlords on account of the political considerations which this gives them. There is, I believe, no-where in Europe, except in England, any instance of the tenant building upon the land of which he had no lease, and trusting that the honour of his landlord would take no advantage of so important an improvement... The law which secures the longest leases against successors of every kind is, so far as I know, peculiar to Great Britain.[28]

These differences were at least several centuries old. Whereas in France in the eighteenth century, Smith had been told that five-sixths of the whole kingdom was still held by some form of older share-cropping agreement, the *métayer*, such tenures 'have been so long in disuse in England that at present I know no English name for them'.[29] These differences in social structure were reflected in the various colonies of France, Spain, England and other European countries. Thus he felt that 'the political institutions of the English colonies have been more favourable to the improvement and cultivation' of the New World than those of Continental countries. One of the central differences was that of alienability of land. In the continental colonies, the land was held as family property, in English colonies as an alienable commodity. Thus he described the differences, whereby in English colonies 'the tenure of the lands, which are all held by free socage, facilitates alienation', whereas in Spanish and Portuguese colonies 'what is called the right of Majorazzo takes place in the succession of all those great estates to which any title of honour is annexed. Such estates go all to one person, and are in effect entailed and unalienable', while in French colonies, 'if any part of an estate, held by the noble tenure of chivalry and homage, is alienated, it is, for a limited time, subject to the right of redemption, either by the heir of the superior or by the heir of the family... which necessarily embarrass alienation'.[30] Thus the English system would tend to create a mass of middling folk, and the Continental systems would re-create the great divide between nobility and peasantry of the homeland.

Smith noted that 'In none of the English colonies is there any hereditary nobility'. There is a difference of esteem, but not of law; 'the descendant of an old colony family is more respected than an upstart of equal merit and fortune: but he is only more respected, and he has no privileges by which he can be troublesome to his neighbours'.[31] Indeed, he argues, it is a feature of the commercial states of which both old and new England were examples, that 'riches . . . very seldom remain long in the same family'.[32] The 'common law of England, indeed, is said to abhor perpetuities' and hence entails were in England 'more restricted than in any other European monarchy'.[33]

Smith's picture of eighteenth-century England and New England is of modern commercial societies. The empire was created to provide buyers for English goods. 'To found a great empire for the sole purpose of raising up a people of customers, may at first sight appear a project fit only for a nation of shopkeepers. It is, however, a project altogether unfit for a nation of shopkeepers; but extremely fit for a nation whose government is influenced by shopkeepers.'[34] Smith assumes that such a mentality is very ancient.

Another part of the virtuous circle which Smith detected was that countries which were growing wealthier could afford greater taxes. 'Easy taxes' were one of his *desiderata* of course, but most civilizations had experienced the reverse; as they became wealthier, the separation of the classes grew, defence became more expensive and that condition which he had noted in China of a vast mass of miserably poor, with heavy rents and taxes, and a small group of very rich, tended to occur. He advocated the reverse. His first principle of taxation was equality. 'The subjects of every state ought to contribute towards the support of the government, as nearly as possible, in proportion to their respective abilities; that is, in proportion to the revenue which they respectively enjoy under the protection of the state.'[35] Secondly the taxation must be certain – that is to say predictable and not arbitrary. 'The time of payment, the manner of payment, the quantity to be paid, ought all to be clear and plain to the contributor, and to every other person.'[36] The arbitrary power of tax gatherers was disastrous. Thirdly, 'Every tax ought to be levied at the time, or in the manner, in which it is most likely to be convenient for the contributor to pay it.'[37] Finally, it should be economically collected, as little as possible being siphoned off in the collection. Here he was describing a world not only of 'easy' taxes, but of a form of taxation to which the Dutch and English were accustomed, but certainly not those living in almost every other agrarian civilization in history. The powerful middle class and weak aristocratic interests were, of course, one of the bulwarks against the normal tendency towards unequal, unpredictable, inconvenient and uneconomical methods.

Smith had developed an aversion to unfair and arbitrary taxation early in his writings. In his lectures he had pointed out 'Whatever policy tends to raise the market price above the naturall one diminishes publick opulence and naturall wea<l>th of the state... Hence it is evident that all taxes on industry must diminish the national opulence as they raise the market price of the commodities.'[38] Yet the merchants were usually too weak to be able to do anything about it. 'All taxes upon exportation and importation of goods also hinder commerce. Merchants at first were in so contemptible a state that the law, as it were, abandoned them, and it was no matter what they obliged them to pay.'[39] All 'fiddling' with the natural order of things was unhelpful, in either a negative or positive way. 'Whatever breaks this naturall balance by giving either an extraordinary discouragement by taxes and duties, or [by] an extraordinary encouragement by bounties or otherwise, tends to hurt the naturall opulence.'[40]

The security and fairness of the tax system was one consequence of the stability of the political system. Smith was very aware that random violence, whether of war, civil war, feuding or even arbitrary justice, would stifle tendencies towards commercial activity. Thus he believed that 'A man must be perfectly crazy who, where there is tolerable security, does not employ all the stock which he commands, whether it be his own or borrowed of other people, in some one or other of those three ways.'[41] On the other hand, capital would become frozen during political insecurity.

> In those unfortunate countries, indeed, where men are continually afraid of the violence of their superiors, they frequently bury and conceal a great part of their stock, in order to have it always at hand to carry with them to some place of safety, in case of their being threatened with any of those disasters to which they consider themselves as at all times exposed. This is said to be a common practice in Turkey, in Indostan, and, I believe, in most other governments of Asia. It seems to have been a common practice among our ancestors during the violence of the feudal government.[42]

Thus the development of 'opulence' depended on the building of a whole infrastructure of legal and quasi-legal protection. Contracts must be binding and enforceable. 'Another thing which greatly retarded commerce was the imperfection of the law with regard to contracts, which were the last species of rights that sustained action, for originaly the law gave no redress for any but those concluded on the spot.'[43] Tenure must be protected. 'As the tenants were continualy in danger of being turned out, they had no motive to improve the ground. This takes place to this day in every country of Europe except Brittain.'[44] Monopolies must be broken down. 'All monopolies and exclusive priviledges of corporations, for whatever good ends they were at first instituted, have the same bad effect.'[45]

One form of constraint on freedom which he noted, particularly in early studies, came not from the State but from the family. Smith explained the growing concern with property in the development from hunter-gatherers to settled pastoralists. 'In the age of hunters there could be no room for succession as there was no property. Any small things as bows, quiver etc. were buried along with the deceased; they were too inconsiderable to be left to an heir. – In the age of shepherds, when property was greatly extended, the goods the deceased had been possessed of were too valuable to be all buried along with him.'[46] Once such valuable property occurred it tended to belong to the kinship group. In the 'age of shepherds', the respect for the family and blood line was particularly strong. 'We see many instances of the vast respect paid to descent amongst the Tartars and Arabs. Every one of these can trace themselves, at least they pretend to do so, as far back as Abraham.'[47]

Yet, as he had himself witnessed in the Scottish Highlands, as the clanship system of the 'shepherds' gave way to the new commercialized economy, the power of kinship declined. 'Regard for remote relations becomes in every country less and less, according as this state of civilization has been longer and more completely established.'[48] Although the loss in martial spirit and warmth was to be lamented, this did increase the opportunities for the individual to keep the fruit of his or her own labour and hence encourage industriousness.

The extreme form of the break with the family could be seen in the spread of the use of last wills and testaments, and the possibility of exclusion of certain family members from the inheritance. Smith was puzzled by the emergence of this means of disposal. 'There is no point more difficult to account for than the right we conceive men to have to dispose of their goods after their death.'[49] He continued to wonder 'how is it that a man comes to have a power of disposing as he pleases of his goods after his death. What obligation is the community under to observe the directions he made concerning his goods now when he can have no will, nor is supposed to have any knowledge of the matter.'[50] He realized that 'In the savage nations of Asia and Africa testamentary succession is unknown; the succession is intirely settled; a man's estate goes always to his nearest male relations, without his having the power of disposing, by any deed to take place after his death, of the smallest subject.'[51]

Yet something very odd and 'individualistic' had emerged in the medieval west, sometime in the period between the collapse of the Roman Empire and the tenth century. 'The German nations which overran Europe had no notion of testamentary succession; every inheritance was divided amongst the children; the only people amongst them (after the introduction of Christianity) who had any such idea were the clergy.'[52] It was thus natural for Smith to argue, as others have done since, that it was the Christian clergy who introduced this device in order to encourage people to leave their property to the church.[53] 'As the clergy were the introducers of testamentary succession, so they were reckond the most proper persons to judge of it, as being best skilled...'[54] It was now possible for wealth to accumulate by the selection of heirs and bypassing the rights of the family at large.

This was a general feature, which was made even more powerful in western Europe, and particularly England, through an institution of which Smith clearly disapproved, namely primogeniture. He argued that the right of individual inheritance by the first born male was a technique introduced to overcome the 'independency of the great allodial estates, and the inconveniencies attending divisions of such estates'.[55] Thus 'this method of succession, so contrary to nature, to reason, and to justice, was occasioned by the nature of feudall government...'[56]

The other form of embedded, birth-given, control over the individual was slavery. Here again Smith noted a progression from almost universal slavery to its gradual elimination. This was both a cause and a consequence of economic growth. He noted first that even in the 1760s, slavery was very widespread. 'We are apt to imagine that slavery is entirely abolished at this time, without considering that this is the case in only a small part of Europe; not remembering that all over Moscovy and all the eastern parts of Europe, and the whole of Asia, that is, from Bohemia to the Indian Ocean, all over Africa, and the greatest part of America, it is still in use.'⁵⁷ He noted a paradox: that slavery became increasingly unacceptable as 'wealth' and equality developed more generally. 'We may observe here that the state of slavery is a much more tolerable one in <a> poor and barbarous people than in a rich and polished one.'⁵⁸

The real puzzle was why an institution which was based on a powerful human drive came to be abolished at all.

> Slavery therefore has been universall in the beginnings of society, and the love of dominion and authority over others will probably make it perpetuall. The circumstances which have made slavery be abolished in the corner of Europe in which it now is are peculiar to it, and which happening to concurr at the same time have brought about that change.⁵⁹

His answer seems to have been that slavery went against the interests and ethics of the King and the Church, both of which had an interest in direct, free, relations with all the citizens or believers in a country. As the power of King and Church grew, so slavery was abolished. 'In Scotland, England, the authority of the king and of the church have been both very great; slavery has of consequence been abolished...'⁶⁰

*

It is often alleged that Smith advocated a weak state. This is a half-truth. In fact what he suggested was that the State should both be strong, as a defence against sectional interests, but also not interfere too much. Ideally the State should be like a referee or umpire – able to punish or even expel, but not actually involved in the everyday contests and exchanges that led to wealth creation.

He believed that it had been private activities and not state interference which had led to the growth of England's wealth over time.

> In the midst of all the exactions of government, this capital has been silently and gradually accumulated by the private frugality and good conduct of individuals, by their universal, continual, and uninterrupted effort to better their own condition. It is this effort, protected by law and allowed by liberty to exert itself in the manner that is most advantageous, which has maintained

the progress of England towards opulence and improvement in almost all former times, and which, it is to be hoped, will do so in all future times.⁶¹

He believed that the ideal situation would be that 'Every man, as long as he does not violate the laws of justice, is left perfectly free to pursue his own interest his own way, and to bring both his industry and capital into competition with those of any other man, or order of men.'⁶² In order to effect and support this system of 'natural liberty'

> the sovereign has only three duties to attend to; three duties of great importance, indeed, but plain and intelligible to common understandings: first, the duty of protecting the society from the violence and invasion of other independent societies; secondly, the duty of protecting, as far as possible, every member of the society from the injustice or oppression of every other member of it, or the duty of establishing an exact administration of justice; and, thirdly, the duty of erecting and maintaining certain public works and certain public institutions, which it can never be for the interest of any individual, or small number of individuals, to erect and maintain; because the profit could never repay the expenc to any individual or small number of individuals, though it may frequently do much more than repay it to a great society.⁶³

Thus Smith realized that the duties were 'of great importance', but they were specific and limited, and included the provision of public, utilities and infrastructure, and a public system of justice.

Smith likewise saw both the merits but also the dangerous absolutist tendencies of organized religion. In an interesting but little-quoted chapter on the 'Institutions for Religious Instruction' he gave a brief account, no doubt heavily influenced by the views of his friend David Hume, of the dangers and advantages of religious enthusiasm.

He noted the danger of politicians taking sides in sectarian squabbles, summarizing his argument in the heading 'If politics had never called in the aid of religion, sects would have been so numerous that they would have learnt to tolerate each other.'⁶⁴ He pointed to the good example of Pennsylvania, where though the Quakers were the most numerous, 'the law in reality favours no one sect more than another, and it is there said to have been productive of this philosophical good temper and moderation'.⁶ He saw that tolerance developed out of the productive balance and tension of different religious positions. Citing Hume's ideas almost verbatim, he wrote:

> In every civilized society, in every society where the distinction of ranks has once been completely established, there have been always two different schemes or systems of morality current at the same time; of which the one may be called the strict or austere; the other the liberal, or, if you will, the loose system. The former is generally admired and revered by the common people: the latter is commonly more esteemed and adopted by what are called people of fashion.⁶⁶

Religious sects, he argued, usually began with the austere, puritanical, position of the country people. They may take this to extreme lengths so that 'in small religious sects morals are regular and orderly and even disagreeably rigorous and unsocial'.⁶⁷ This puritanical attitude can be ameliorated by encouraging such sectarians to broaden their minds with science and philosophy, painting, poetry, music, dancing and such things.

Smith then considers the dangers of an Established Church, which tends again to become too powerful. Combined with the growing wealth of the State this made it 'exceedingly formidable'.⁶⁸ The extreme example of this tendency was, of course, the Papacy,

> In the state in which things were through the greater part of Europe during the tenth, eleventh, twelfth, and thirteenth centuries, and for some time both before and after that period, the constitution of the church of Rome may be considered as the most formidable combination that ever was formed against the authority and security of civil government, as well as against the liberty, reason, and happiness of mankind, which can flourish only where civil government is able to protect them.⁶⁹

What then brought down this great and increasing power, as potent a threat as the feudal lords? Smith suggests the same force as before, namely the growth of commercial wealth, and in exactly the same way. In other words, it was not destroyed from outside, but corrupted by greed from inside.

> The gradual improvements of arts, manufactures, and commerce, the same causes which destroyed the power of the great barons, destroyed in the same manner, through the greater part of Europe, the whole temporal power of the clergy. In the produce of arts, manufacturers, and commerce, the clergy, like the great barons, found something for which they could exchange their rude produce, and thereby discovered the means of spending their whole revenues upon their own persons, without giving any considerable share of them to other people. Their charity became gradually less extensive, their hospitality less liberal or less profuse. Their retainers became consequently less numerous, and by degrees dwindled away altogether. The clergy too, like the great barons, wished to get a better rent from their landed estates, in order to spend it, in the same manner, upon the gratification of their own private vanity and folly. But this increase of rent could be got only by granting leases to their tenants, who thereby became in a great measure independent of them. The ties of interest, which bound the inferior ranks of people to the clergy, were in this manner gradually broken and dissolved.[70]

This internal corruption had weakened the established churches well before the Reformation. But that movement was the final blow. The enthusiasm of the Reformers was supported by the puritanical zeal of ordinary people, and thus 'enabled sovereigns on bad terms with Rome to overturn the Church with ease'.[71]

The form of government in England, whereby the Lutherans formed a weak link with the Crown, 'was from the beginning favourable to peace and good order, and to submission to the civil sovereign'.[72] In Scotland the Calvinist system had been less successful because the 'Election by the people gave rise to great disorders', with a fanatical clergy and factions and controversies.[73] This period of disorder was ended by the various early eighteenth century acts which helped to diminish the factionalism. Thus by his own time, Smith could comment that 'There is scarce perhaps to be found any where in Europe a more learned, decent, independent, and respectable set of men, than the greater part of the presbyterian clergy of Holland, Geneva, Switzerland and Scotland.'[74] Thus Church and State were reasonably balanced, and neither was either willing or able to halt the progress towards the general improvements in trade, manufacture and wealth, in Smith's meaning of that word.

8
From Predation to Production

Living within a few miles of the Highland Line, and having narrowly avoided the forays of the Scots clan-based army in 1745, Adam Smith was deeply aware of how fragile and original was the kind of commercial order he saw in England. Thus when he wrote that all that was needed was 'peace, easy taxation and a tolerable administration of justice', he not only selected three political conditions but must have been fully aware that such conditions were incredibly difficult to attain. He was not making a statement about how *easy* the 'natural course' of opulence was, but how difficult. How then had these conditions emerged, in particular in England? This is one of the trickiest of questions, the relations between power and wealth. The powers of predation were bound to be stronger and more desirable than the powers of production. So how did wealth creation ever continue in any sustained and prolonged way?

The question could be put in the form: how did violence gradually ebb away? Smith has several lines of argument to explain this, but the central, and somewhat circular one, is that people were gradually 'civilized' by increasing wealth – or as Samuel Johnson put it, 'There are few ways in which a man can be more innocently employed than in getting money.'[1] Here is the argument, incorporating a certain amount of questionable history, based on Smith's knowledge of England and France.

At the start of the period with which Smith was concerned, the world approximated Marc Bloch's 'dissolution of the State' feudalism, with powerful lords and their bands of retainers and castles, as Smith must have witnessed in his youth in the Highlands. The centre was very weak. The king in those 'ancient times' was little more than the greatest proprietor in his dominions, to whom, for the sake of common defence against their common enemies, the other great proprietors paid certain respects. To have enforced payment of a small debt within the lands of a great proprietor, where all the inhabitants were armed and accustomed to stand by one another, would have cost the king, had he attempted it by his own authority, almost the same effort as to extinguish a civil war. He was, therefore, obliged to abandon the administration of justice through the greater part of the country, to those who were capable of administering it; and for the same reason to leave the command of the country militia to those whom that militia would obey.[2]

Thus the great proprietors, Smith thought, had the power to raise troops, execute justice and so on before and after the Norman Conquest of England. Gradually the imposition of feudal law after the twelfth century led to some reining in of the over-mighty barons. 'The introduction of the feudal law, so far from extending, may be regarded as an attempt to moderate the authority of the great allodial lords. It established a regular subordination, accompanied with a long train of services and duties, from the king down to the smallest proprietor.'[3] Yet even after the introduction of feudal subordination, he believed, 'the king was as incapable of restraining the violence of the great lords as before. They still continued to make war according to their own discretion, almost continually upon one another, and very frequently upon the king; and the open country still continued to be a scene of violence, rapine, and disorder.'[4]

So what turned the tide of violence if it was not the political system? Here, returning to the theme of the civilizing effect of commerce, Smith brings forward his explanation.

But what all the violence of the feudal institutions could never have effected, the silent and insensible operation of foreign commerce and manufactures gradually brought about. These gradually furnished the great proprietors with something for which they could exchange the whole surplus produce of their lands, and which they could consume themselves without sharing it either with tenants or retainers. All for ourselves, and nothing for other people, seems, in every age of the world, to have been the vile maxim of the masters of mankind.[5]

Here again Smith was writing exactly about his own, post-Culloden experience as he watched the Scottish clan lords dismiss their followers and turn their lands over to sheep and other more profitable commodities with which they could raise the cash to leave the Highlands and live in the cities of southern Scotland or England.

> In a country where there is no foreign commerce, nor any of the finer manufactures, a man of ten thousand a year cannot well employ his revenue in any other way than in maintaining, perhaps, a thousand families, who are all of them necessarily at his command. In the present state of Europe, a man of ten thousand a year can spend his whole revenue, and he generally does so, without directly maintaining twenty people, or being able to command more than ten footmen not worth the commanding.[6]

Thus the retainers were sacked and the lords became consumers in a commercial society.

Likewise the tenants were reduced.

> Farms were enlarged, and the occupiers of land, notwithstanding the complaints of depopulation, reduced to the number necessary for cultivating it, according to the imperfect state of cultivation and improvement in those times. By the removal of the unnecessary mouths, and by exacting from the farmer the full value of the farm, a greater surplus, or what is the same thing, the price of a greater surplus, was obtained for the proprietor, which the merchants and manufacturers soon furnished him with a method of spending upon his own person in the same manner as he had done the rest.[7]

Although Smith does not explicitly say so, he is both describing what he saw happening before him in the Highland clearances and projecting it backwards as a model for what he thought must have happened in England in the later Middle Ages as a feudal society gave way to a commercial one.

This revolution was an unintended and accidental event. No one was aware of what was happening, partly because the change happened in England over a long period. Hence, as the marginal heading put it, 'A revolution was thus insensibly brought about'. What this revolution was, and its accidental nature, is summarized by Smith thus.

> A revolution of the greatest importance to the public happiness, was in this manner brought about by two different orders of people, who had not the least intention to serve the public. To gratify the most childish vanity was the sole motive of the great proprietors. The merchants and artificers, much less ridiculous, acted merely from a view to their own interest, and in pursuit of their own pedlar principle of turning a penny wherever a penny was to be got. Neither of them had either knowledge or foresight of that great revolution which the folly of the one, and the industry of the other, was gradually bringing about.[8]

All this was, of course, topsy-turvy and hence progress was much slower than it should be. For 'It is thus that through the greater part of Europe the commerce and manufactures of cities, instead of being the effect, have been the cause and occasion of the improvement and cultivation of the country.'[9] This 'being contrary to the natural course of things', was 'necessarily both slow and uncertain'.[10] Yet it had happened – just. What should really happen was shown by developments in North America. 'Compare the slow progress of those European countries of which the wealth depends very much upon their commerce and manufactures, with the rapid advances of our North American colonies, of which the wealth is founded altogether in agriculture.'[11]

The process was complex and contained many feedback loops. As the wealth increased, so the legal framework upon which Smith believed flourishing commercialism must be based grew stronger. A particularly important factor was the increasing separation of the executive and the judiciary, in other words an impartial judiciary which will protect citizens from that arbitrary arrogance of power to be found in most agrarian and totalitarian empires. Here again, Smith thought the key was growing wealth.

The idea of the importance of 'tolerable administration of justice' as one of the keys to capitalism was expanded by Smith.

> When the judicial is united to the executive power, it is scarce possible that justice should not frequently be sacrificed to, what is vulgarly called, politics. The persons entrusted with the great interests of the state may, even without any corrupt views, sometimes imagine it necessary to sacrifice to those interests the rights of private man. But upon the impartial administration of justice depends the liberty of every individual, the sense which he has

of his own security. In order to make every individual feel himself perfectly secure in the possession of every right which belongs to him, it is not only necessary that the judicial should be separated from the executive power, but that it should be rendered as much as possible independent of that power. The judge should not be liable to be removed from his office according to the caprice of that power. The regular payment of his salary should not depend upon the good-will, or even upon the good oeconomy of that power.[12]

This crucial change came about because, as certain nations became wealthier, the amount of business increased, and hence the famous division of labour applied here also.

The separation of the judicial from the executive power seems originally to have arisen from the increasing business of the society, in consequence of its increasing improvement. The administration of justice became so laborious and so complicated a duty as to require the undivided attention of the persons to whom it was entrusted. The person entrusted with the executive power, not having leisure to attend to the decision of private causes himself, a deputy was appointed to decide them in his stead.[13]

Smith believed that this had happened in the Roman empire, and again in medieval and early modern England. It is an ingenious idea, but does not fully accord with what happened in the Turkish, Habsburg, Russian or Chinese empires. Obviously, growing wealth is only part of the answer.

Another ingenious circularity lies in the effect of growing wealth on the propensity to violence. Basically the argument is that people are too busy to be violent, and find it more convenient to follow the principle of the division of labour and buy off the threat of violence by hiring others to police their borders. This leads to an overall decline in warlike feelings on the part of the majority of the population. Smith put it in an evolutionary way thus.

A shepherd has a great deal of leisure; a husbandman, in the rude state of husbandry, has some; an artificer or manufacturer has none at all. The first may, without any loss, employ a great deal of his time in martial exercises; the second may employ some part of it; but the last cannot employ a single hour in them without some loss, and his attention to his own interest naturally leads him to neglect them altogether. Those improvements in husbandry too, which the progress of arts and manufactures necessarily introduces, leave the husbandman as little leisure as the artificer. Military exercises come to be as much neglected by the inhabitants of the country as by those of the town, and the great body of the people becomes altogether unwarlike.[14]

Smith's thesis was

confirmed by universal experience. In the year 1745 four or 5 thousand naked unarmed Highlanders took possession of the improved parts of this country without any opposition from the unwarlike inhabitants. They penetrated into England and alarmed the whole nation, and had they not been opposed... they would have seized the throne with little difficulty. 200 years ago such an attempt would have rouzed the spirit of the nation.[15]

This was specifically the result of the effects of the division of labour and commerce. 'Another bad effect of commerce is that it sinks the courage of mankind, and tends to extinguish martial spirit. In all commercial countries the division of labour is infinite, and every ones thoughts are employed about one particular thing.'[16]

Smith's ambivalence about this process is shown even more clearly when he wrote: 'The defence of the country is therefore committed to a certain sett of men who have nothing else ado; and among the bulk of the people military courage diminishes. By having their minds constantly employed on the arts of luxury, they grow effeminate and dastardly.'[17] This was a 'bad' effect because it meant that almost automatically, as a nation became richer, it became a prey to others. They were both more attractive and more vulnerable. It was a situation which was obvious to Smith.

That wealth, at the same time, which always follows the improvements of agriculture and manufactures, and which in reality is no more than the accumulated produce of those improvements, provokes the invasion of all their neighbours. An industrious, and upon that account a wealthy nation, is of all nations the most likely to be attacked; and unless the state takes some new measures for the public defence, the natural habits of the people render them altogether incapable of defending themselves.[18]

He expanded this theme at length in his lectures on jurisprudence, with particular reference to the Roman empire. He showed how

> when the whole people comes to be employed in peacefull and laborious arts, 1 out of 100 only can go, that is, about 1000, which would be no more than a poor city guard and could do nothing against an enemy; nor even 4 or 5000. So that the very duration of the state and the improvements naturally going on at that time, every one applying himself to some usefull art, and commerce, the attendant on all these, necessarily undo the strength and cause the power to vanish of such a state till it be swallowed up by some neighbouring state.[19]

This forced a state to turn to using either the dregs of society, or paid mercenaries.

> Thus then when arts were improved, those who in the early times of the state had alone been trusted would not now go out, and those who before had never engaged in battle were the only persons who made up the armies, as the proletarii or lowest class did in the later periods of Rome. The armies are diminished in number but still more in force. This effect commerce and arts had on all the states of Greece. We see Demosthenes urging them to go out to battle themselves, instead of their mercenaries which their army then consisted of; nor of these were there any considerable number. Whenever therefore arts and commerce engage the citizens, either as artizans or as master trades men, the strength and force of the city must be very much diminished.[20]

This was the fate of all small enclaves like the Greek city states. 'All states of this sort would therefore naturally come to ruin, its power being diminished by the introduction of arts and commerce, and its territory, and even its very being, being held on a very slender tack after the military art was brought to tollerable perfection, as it had nothing to hope for when once defeated in the field.'[21] There was a vicious circle. The very force that led to wealth led to ruin. 'Here improvement in arts and cultivation unfit the people from going to war, so that the streng<th> is greatly diminished and it falls a sacrifice to some of its neighbours. This was the case of most of the republicks of Greece. Athens in its later time could not send out the 5th part of what it formerly did.'[22] A classic case was also provided by the fate of the Italian city states. 'The Italian republicks in the same manner paid subsidies to some of the neighbouring chiefs who engaged to bring 10 000 or 5000 horse, which were then chiefly in request, for their protection. Every small state had some of these in their pay. This soon brought on their ruin.'[23]

Their weakness was increased by changes in military technology. With improved weaponry it became impossible to defend a city, for

> there is an other improvement which greatly diminishes the strength and security of such a state; I mean the improvement of the military art. The taking of cities was at fir<s>t a prodigious operation which employed a very long time, and was never accomplished but by stratagem or blockade, as was the case at the Trojan war. A small town with a strong wall could hold out very well against its enemies.[24]

If such a city state tried an alternative tactic, that is imperial expansion, it ran into the problems which Montesquieu had outlined. Smith believed that the lust for domination and power was very strong.

> The Love of domination and authority over others, which I am afraid is naturall to mankind, a certain desire of having others below one, and the pleasure it gives one to have some persons whom he can order to do his work rather than be obliged to persuade others to bargain with him, will for ever hinder this from taking place.[25]

As Rome expanded, so the army and its generals became more powerful.

> Of all the republicks we know, Rome alone made any extensive conquests, and became thus in danger from its armies under the victorious leaders. But the same thing was feared and must have happend at Carthage had the project of Hanniball succeded, and he made himself master of Italy.[26]

The classic instance was, of course, Caesar. The behaviour of the Senate in Rome

> affronted Caesar; he had recourse to his army who willingly joind him, and by repeated victories he became Dictator for ever. The remains of the same victorious army afterwards set Antony and Augustus, and at last Augustus alone, on the throne. And the same will be the case in all conquering republicks where ever a mercenary army at the disposall of the generall is in use.[27]

How could a country or confederacy of states escape from this vicious circle? They could not rely on international law.

> In war, not only what are called the laws of nations are frequently violated, without bringing (among his own fellow-citizens, whose judgments he only regards) any considerable dishonour upon the violator; but those laws themselves are, the greater part of them, laid down with very little regard to the plainest and most obvious rules of justice.[28]

Smith's only hope was that the incessant warfare would become milder. Earlier wars had been fought out of wanderlust and in a pure predatory fashion – the search for booty; modern commercial nations fought in order to secure or increase their territory. 'A polished nation never undertakes any such expeditions. It never makes war but with a design to enlarge or protect its territory; but these people make war either with design to leave their own habitations in search of better, or to carry off booty.'[29] This change of motive reduced the destructive element in war. 'The same policey which makes us not so apt to go to war makes us also more favourable than formerly after an entire conquest. Anciently an enemy forfeited all his possessions, and was disposed of at the pleasure of the conquerors. It was on this account that the Romans had often to people a country anew and send out colonies. It is not so now. A conquered country in a manner only changes masters.'[30] The change was only at the very top. 'They may be subjected to new taxes and other regulations, but need no new people. The conqueror generally allows them the possession of their religion and laws, which is a practice much better than the ancient.'[31]

This milder form of conquest was accompanied by a less bloody form of warfare, thanks to modern weapons. Adam Smith knew the difference between Highlanders armed with claymores, and English troops with their muskets. 'Modern armies too are less irritated at one another because fire arms keep them at a greater distance. When they always fought sword in hand their rage and fury were raised to the highest pitch, and as they were mixed with one another the slaughter was vastly greater.'[32] All of this change, however, depended on neighbouring societies all being 'enlightened' and commercial. Until the fifteenth century or so, eastern Europe had been the prey of powerful Mongol armies. Only recently had the new form of 'commercial' warfare become dominant – and destructive though it was, it was less disastrous than that which China and western European nations had faced for thousands of years. Yet Smith was still left with the puzzle of that crucial movement from small, vulnerable, commercial city-states, to large but not too large nation-states which somehow combined commercial affluence with the power to defend themselves. The answer lay in English history.

*

We saw that Montesquieu singled out England as his extreme case of liberty and Smith was to do the same, also emphasizing its wealth. The problem was to account for its success, for, as Campbell and Skinner note, 'Smith believed that England was really a special case, and that she alone had escaped from absolutism.'[33] In his *Lectures on Jurisprudence* he gave a narrative of how this had happened.

He believed that after the collapse of the Roman Empire two forms of government succeeded each other. At first from about 400 to 800 AD there was a form of 'allodial government' where 'the lords held their lands of no one, but possessed them as their own property'. Then 'the feudall government arose about 400 years afterwards, about the 9th century'.[34] This emerged largely out of a new power relationship between the lords and the king. 'When any of the great allodial lords was in danger of being oppressed by his neighbours, he called for protection from the king against them. This he could not obtain without some consideration he should perform to him. A rude and barbarous people who do not see far are very ready to make concessions for a temporary advantage.'[35] The new arrangement was as follows. 'For these considerations the king gave up all his demesne lands, and the great allodiall lords their estates, to be held as feuda, which before had been held as munera. A tenent who held a feu was very near as good as property. He held it for himself and his heirs for ever. The lord had the dominium directum, but he had the dominium utile which <?was> the principle and most beneficial part of property.'[36] The chain of feudal links was thus set up. 'Those services secured his protection, and in this manner the inferior allodial lords came to hold of the great ones, and these again of the king; and the whole thus held of him either mediately or immediately, and the king was conceived to have the dominium directum of all the lands in the kingdom.'[37] This was a change which 'happened in the whole of Europe about the 9, 10, and 11st centuries'.[38]

All the lands 'fell under the immediate jurisdiction of the lords or of the king, who administered judgment in them either by himse<l>f or by judges sent for that purpose'.[39] One side effect of this was to protect the lowest tenants. They avoided the fate of all previous agricultural workers, namely slavery, for though they were 'unfree',

> They were however in a much better condition than the slaves in ancient Greece or Rome. For if the master killed his villain he was liable to a fine; or if he beat him so as that he died within a day he was also liable to a fine; these, tho small priviledges, were very considerable and shewd great superiority of condition if compared with that of the old slaves.[40]

Thus up to about the eleventh century, all of western Europe was fairly uniform. After this Smith began to detect a growing divergence. Over continental Europe the power of the ruler increased. Only in England did this not happen. Everywhere powerful rulers overthrew

> the democraticall part of the constitution and establish an aristocraticall monarchy. This was done in every country excepting England, where the democraticall courts subsisted long after and usually did business; and at this day the county court, tho it has not been used of a long time, is nevertheless still permitted by law.[41]

The nobility, which could have put up a resistance to the absolutist tendency, were crushed, a necessary precursor to liberty, but leading to a dangerous void. 'We see too that this has always been the case; the power of the nobles has allways been brought to ruin before a system of liberty has been established, and this indeed must always be the case. For the nobility are the greatest opposers and oppressors of liberty that we can imagine.'[42]

They were also weakened because of the growing commercial prosperity – a process Smith had observed with the Highland chiefs. Speaking of a typical lord or laird, he wrote,

> When luxury came in, this gave him an opportunity of spending a great deal and he therefore was at pains to extort and squeze high rents from them. This ruind his power over them. They would then tell him that they could not pay such a rent on a precarious chance of possession, but would consent to it if he would give them long possessions of them; which being convenient for both is readily agreed to; and they became still more independent when in the time of Henry 2d these leases came to sustain action at law contra quemcumque possessorem. Thus they lost a man for 10 or 5 sh., which they spent in follies and luxury. The power of the lords in this manner went out, and as this generally happened before the power of the commons had come to any great pitch, an absolute government generally followed.[43]

This was a necessary stage between feudalism and modern liberty.

> Whereas every one is in danger from a petty lords, who had the chief power in the whole kingdom. The people therefore never can have security in person or estate till the nobility have been greatly crushed. Thus therefore the government became absolute, in France, Spain, Portugal, and in England after the fall of the great nobility.[44]

Smith had noted a difference between England and the Continent early on, and continues this theme from time to time up to the later fifteenth century. Thus he argues that the English were often able to gain some freedom and power when their rulers needed money to fight wars abroad.

> The people we see were always most free from their severall burthens when the profits arising from them to the state were most necessary for its support. We see accordingly that those which are most favourable to liberty are those of martiall, conquering, military kings. Edward the 1st and Henry the 4th, the

two most warlike of the English kings, granted greater immunities to the people than any others.'[45]

He explained why this should be so.

> There are severall reasons for this, as 1st, they of all others depended most on the goodwill and favour of their people; they therefore court it greatly by all sorts of concessions which may induce them to join in their enterprizes. Peaceable kings, who have no such occasion for great services or expensive expedition<s>, [and] therefore less courted their love and favour. 2dly, it soon became a rule with the people that they should grant no subsidies till their requests were first granted.[46]

Thus the loss of English interest in their French claims in the fifteenth century, according to Smith, was a disaster. The nobility were weakened in the Wars of the Roses. 'They had been massacred by Edward 4th in his battles with Henry the 6th; afterwards in various insurrections and disputes for the crown.'[47] Under the Tudors they sunk into a form of absolutism.

> In the last lecture I observd how the nobility necessarily fell to ruin as soon as luxury and arts were introduced. Their fall everywhere gave occasion to the absolute power of the king. This was the case even in England. The Tudors are now universally allowed to have been absolute princes. The Parliament at that time, instead of apposing and checking the measures they took to gain and support their absolute power, authorized and supported them. Henry the 7th was altogether an absolute monarch...[48]

This narrative left Smith with a problem. All of Europe was now absolutist. How then in the less than two hundred years between Henry VII and the later seventeenth century had England, alone, shed absolutism? Smith put forward two arguments. The first was the rash activity of Elizabeth I. A thing which

> contributed to the diminution of the kings authority, and to render him still more weak, was that Elizabeth in the end of her reign, forseeing that she was to have no sucessors of her own family, was at great pains to gain the love of the nation, which

she had generally done, and never inclined to lay on taxes which would she knew be complaind of; but she chose rather to sell the demesne lands, which were in her time alltogether alienated. James Ist and Charles had in this manner no revenue, nor had they a standing army by which they could extort any money or have other influence with the people.[49]

Smith explained how absolutism increased over continental Europe. But the

> situation and circumstances of England have been altogether different. It was united at length with Scotland. The dominions were then entirely surrounded by the sea, which was on all hands a boundary from its neighbours. No foreign invasion was therefore much to be dreaded. We see that (excepting some troops brought over in rebellions and very impoliticly as a defence to the kingdom) there has been no foreign invasion since the time of Henry 3d.... The Scots however frequently made incursions upon them, and had they still continued seperate it is probable the English would never have recovered their liberty. The Union however put them out of the danger of invasions. They were therefore under no necessity of keeping up a standing army; they did not see any use of necessity for it.[50]

He contrasted this with the position of continental nations.

> In other countries, as the feudall militia and that of a regular one which followd it wore out, they were under a necessity of establishing a standing army for their defense against their neighbours. The arts and improvement of sciences puts the better sort in such a condition that they will not incline to serve in war. Luxury hinders some and necessary business others.[51]

Thus the English and French diverged through a combination of Elizabeth's profligacy and the security of England's island position after union with Scotland.

> We see in France that Henry the fourth [c. 1600] kept up generally a standing army of betwixt 20 and 30 000 men; this, tho small in comparison of what they now keep up, was reckoned a great force, and it was thought that if France could in time of

peace maintain that number of men it would be able to give law to Europe; and we see it was in fact very powerfull. But Britain had no neighbours which it could fear, being then thought superior to all Europe besides. The revenues of the king being very scanty, and the desmesnes lands, the chief support of the kings, being sold, he had no more money than was necessary to maintain the dignity and grandeur of the court. From all these, it was thought unnecessary as well as inconvenient and useless to establish a standing army.[52]

The result was that when the Civil War was fought, Charles lost. Again, when James II tried to impose his Catholic will, he could not do so. Thus, for largely accidental and fortuitous reasons, England's political history took a different turn.

Smith was not content to leave the story here, however. Following closely some of the arguments of Montesquieu, he tried to outline the institutional structures which now guaranteed the balance of power and the liberty and security of the people of Britain, which was the foundation, as he thought, of their growing wealth and power. There was firstly England's parliament. By the middle of the eighteenth century,

> So far is the king from being able to govern the kingdom without the assistance of Parliament for 15 or 16 years, as Chas. Ist did, that he could not without giving offence to the whole nation by a step which would shock every one, maintain the government for one year without them, as he has no power of levying supplies. In this manner a system of liberty has been established in England before the standing army was introduced; which as it was not the case in other countries, so it has not been ever establishd in them.[53]

The system was now firmly entrenched. 'Liberty thus established has been since confirmed by many Acts of Parliament and clauses of Acts. The system of government now supposes a system of liberty as a foundation. Every one would be shocked at any attempt to alter this system, and such a change would be attended with the greatest difficulties.'[54] The House of Commons was powerful enough to control the royal power and the power of ministers. 'Another article which secures the liberty of the subjects is the power which the Commons have of impeaching the kings ministers of mal-administration, and that tho it had not visibly encroached on liberty.'[55] Furthermore, 'The House of Commons also has the sole judgement in all controverted elections, and is on them very nice and delicate, as their interest leads them to preserve them as free as can be had.'[56]

Yet the Commons themselves were restrained from the corruption of power by periodic elections. 'The frequency of the elections is also a great security for the liberty of the people, as the representative must be carefull to serve his country, at least his constituents, otherwise he will be in danger of losing his place at the next elections.'[57] In summary, 'These laws and established customs render it very difficult and allmost impossible to introduce absolute power of the king without meeting with the strongest opposition imaginable.'[58] All this had occurred through a balance of forces. Smith rejected both Hobbes's and Locke's ideas of the social contract arising from a voluntary agreement. As he explained, he had 'endeavoured to explain to you the origin and something of the progress of government. How it arose, not as some writers imagine from any consent or agreement of a number of persons themselves to submit themselves to such or such regulations, but from the natural progress which men make in society.'[59]

The other great protection, in suggesting which Smith again partly followed Montesquieu, was the English common law tradition. One aspect was a free and independent judiciary, judges who were separate from the royal or even the parliamentary power.

> One security for liberty is that all judges hold their office[r]s for life and are intirely independent of the king. Every one therefore is tried by a free and independent judge, who are als<o> accountable for their conduct. Nothing therefore will influ<en>ce them to act unfairly to the subject, and endang<er> the loss of a profitable office and their reputation also; nothing the king could bestow would be an equivalent. The judge and jury have no dependance on the crown.[60]

He expanded this a little later, by pointing out that the judges themselves were limited in their power.

> I had observed an other thing which greatly confirms the liberty of the subjects in England. – This was the little power of the judges in explaining, altering, or extending or correcting the mean-

ing of the laws, and the great exactness with which they must be observed according to the literall meaning of the words, of which history affords us many instances.⁶¹

Part of this limitation, which prevented yet another danger, that of an arbitrary justice, was due to the healthy rivalry between different courts – providing reasonable competition in justice.

> Another thing which tended to support the liberty of the people and render the proceedings in the courts very exact, was the rivalship which arose betwixt them. The Court of Kings Bench, being superior to the Court of Com. Pleas and having causes frequently transferred to them from that court, came to take upon it to judge in civill causes as well as in criminall ones, not only after a writ of error had been issued out but even immediately before they had passed thro the Common Pleas.⁶²

Smith had already drawn attention to the independence of jurors, but he elaborated this further as a key protection of the citizen against the State, and also against the power of judges.

> Another thing which curbs the power of the judge is that all causes must be try'd with regard to the fact by a jury. The matter of fact is left intirely to their determination. – Jurys are an old institution which formerly were in use over the greater part of the countries in Europe, tho they have now been laid aside in all countries, Britain excepted.⁶³

Great care was taken to maintain their independence and reliability. 'Nothing can be more carefull and exact than the English law in ascertaining the impartiality of the jurers. They must be taken from the county where the persons live, from the neighbourhood of the land if it be a dispute of property, and so in other cases.'⁶⁴

Thus an independent, but not too powerful, judge and an independent jury seemed together 'to be a great security of the liberty of the subject'. As Smith explained to his Scottish students, in England,

> One is tried here by a judge who holds his office for life and is therefore independent and not under the influence of the king, a man of great integrity and knowledge who has been bred to

the law, is often one of the first men in the kingdom, who is also tied down to the strict observance of the law; and the point of fact also determined by a jury of the peers of the person to be tried, who are chosen from your neighbourhood, according to the nature of the suit, all of whom to 13 you have the power of challenging.[65]

The final protection was *Habeas Corpus* , which is 'also a great security against oppression, as by it any one can procure triall at Westminster within 40 days who can afford to transport himself thither'.[66] This prevented arbitrary imprisonment without trial.

Smith's account of the English development is intriguing and scholarly. He clearly knew the literature and wrote with authority. Yet there is something of a contradiction in it. On the one hand his account of political power suggests that England like continental Europe went through an absolutist phase. The difference was that it occurred much later, lasted for a much shorter period (c.1475–1580) and was overturned, whereas the absolutist governments grew ever more powerful in France, Spain, Germany and elsewhere.

On the other hand, on the legal side he gives a sketch of much more continuity and of the preservation of a high degree of protection against the power of the State. He summarizes his finding in this area as follows.

There seems to be no country in which the courts are more under regulation and the authority of the judge more restricted. The form of proceedings as well as the accuracy of the courts depends greatly on their standing. Now the courts of England are by far more regular than those of other countries, as well as more ancient. The courts of England are much more ancient than those of France or Scotland.[67]

It was one of several contradictions in his portrayal of English history which make his account suggestive about the events after about 1600, but less accurate for the earlier period.

*

There are a number of reasons for looking on Adam Smith as an optimist. He believed, like Pope, that whoever stood behind the visible world, had intended mankind to be happy. 'The happiness of mankind, as well as of all other rational creatures, seems to have been the original purpose intended by the Author of Nature when he brought them into existence.'[68] In general such an 'Author' had been successful. 'Take the whole earth at an average, for one man who suffers pain or misery, you will find twenty in prosperity and joy, or at least in tolerable circumstances.'[69]

This view was confirmed by the history of the fairly recent past. Firstly, he could see that progress had been made over most of Europe since the fifteenth century and that his own Scotland was, in parts, becoming very much richer. Violence was on the retreat. As Eric Roll summarizes his Enlightenment optimism here, 'Fundamentally, he, like most later liberal philosophers, was an optimist. The social evils which he saw around him he ascribed to past mistakes of government... Smith's whole work implied great faith in the possibility of freeing the state from the incubus of individual or class influence. Once this emancipation was achieved the natural harmony would be manifest to all.'[70] The very basis of his work was the belief both that an 'Inquiry into the Nature and Causes of the Wealth of Nations' was possible and that, having found such causes, nations could take appropriate and remedial action.

A particular cause for optimism was his belief that the balance between production and predation had changed. A cause for quiet confidence was the fact that wealth and general virtue were connected. As Dugald Stewart noted of his findings, 'the most wealthy nations are those where the people are the most laborious, and where they enjoy the greatest degree of liberty'.[71] They were also characteristically more equal societies, with a large and mobile middling group and the decline of serfdom and aristocracy. The trouble was that when such societies had emerged anywhere else before, as in the Italian city-states, they had quickly been destroyed by envious neighbours. The forces of destruction or predation had always been too strong for the pockets of wealth to resist.

As Adam Smith reflected on the last major contest in British soil between predation and production, the clash between the warlike clans and the mercenary army of the English at Culloden in 1745, it must have been very obvious that the balance had shifted. How and why this had happened, being strongly related to technological changes, is outlined by Smith as follows.

He noted that over time the cost of defence increased as nations became wealthier.

> The first duty of the sovereign, therefore, that of defending the society from the violence and injustice of other independent societies, grows gradually more and more expensive, as the society advances in civilization. The military force of the society, which originally cost the sovereign no expence either in time of peace or in time of war, must, in the progress of improvement, first be maintained by him in time of war, and afterwards even in time of peace.[72]

This cost was increased still further by modern weapons technology. The battle of Culloden was decided by fire power, by superior technology, more than anything else.

> The great change introduced into the art of war by the invention of fire-arms, has enhanced still further both the expence of exercising and disciplining any particular number of soldiers in time of peace, and that of employing them in time of war. Both their arms and their ammunition are become more expensive. A musquet is a more expensive machine than a javelin or a bow and arrows; a cannon or a mortar than a balista or a catapulta. The powder, which is spent in a modern review, is lost irrecoverably, and occasions a very considerable expence.[73]

The effect of this was to favour the rich, rather than those who had previously ruled the earth, the warlike. Rich shopkeepers could now easily defeat poor Highlanders.

> In modern war the great expence of fire-arms gives an evident advantage to the nation which can best afford that expence; and consequently, to an opulent and civilized, over a poor and barbarous nation. In ancient times the opulent and civilized found it difficult to defend themselves against the poor and barbarous nations. In modern times the poor and barbarous find it difficult to defend themselves against the opulent and civilized. The invention of firearms, an invention which at first sight appears to be so pernicious, is certainly favourable both to the permanency and to the extension of civilization.[74]

Thus Smith believed that one of the negative feedback mechanisms which had constantly operated in the past, bringing down the Roman empire, leaving civilizations vulnerable to Mongol invasions, even keeping his native Scotland in thrall, had at last been overcome. Wealth and military power were for the first time united with liberty and equality.

This is the optimistic side. Yet at another level, Smith, like the successor classical economists Malthus and Ricardo, was a pessimist – and for exactly the same reasons. As he looked around him he saw that progress was possible – up to a limit, but then seemed to hit some invisible barrier or ceiling. China was the great example: that mighty civilization, wealthier than Europe, seemed to have been 'stationary' since the time of Marco Polo. India was not 'progressing'. The shape of things to come was shown by Holland, which had been 'stationary', if not declining, for nearly 100 years. France, previously very wealthy, had also been 'stationary' for about a hundred years or so. Italy had only recovered the level of her pre-1500 eminence. Spain and Portugal were 'going backwards'. Only England, still with some way to reach Holland's level, and tiny Scotland and the under-populated spaces of North America, were progressing rapidly.

E.A. Wrigley has summarized an aspect of Smith's pessimism; 'his view of the prospects of growth in general induced him to discount the possibility of a prolonged or substantial improvement in real wages, and to fear that the last state of the labourer would prove to be worse than the first, a view that was reinforced by his anticipation of some of the arguments to which Malthus was later to give the classic formulation.'[75] Smith could not see what would in fact happen. '... Smith himself was unaware of the immense changes already in train when the *Wealth of Nations* was written. Indeed, the implications of the arguments he used would rule out the possibility of rapid and sustained economic growth. The great revolution of which he wrote was an economic revolution ... but it was not an *industrial revolution* as that term has come to be used.'[76] What he accurately described was a closed system which, in the case of most of the European countries and China, had reached the limits of possible progress. He had observed a fact which Wrigley endorses, which is that 'In their essential nature traditional economies were negative feedback systems. At some point the growth process itself provoked changes which caused growth to decelerate and grind to a halt. Success in a particular round of growth implied difficulty at a later stage.'[77]

Smith gave three major reasons why there was no possibility of continuous long-term growth and why a country such as Holland had just about reached the limits. One of these was that the rate of profit would continually fall. This mechanism was demonstrated by the history of Holland.

> In a country which had acquired its full complement of riches, where in every particular branch of business there was the greatest quantity of stock that could be employed in it, as the ordinary rate of clear profit would be very small, so the usual market rate of interest which could be afforded out of it, would be so low as to render it impossible for any but the very wealthiest people to live upon the interest of their money. All people of small or middling fortunes would be obliged to superintend themselves the employment of their own stocks. It would be necessary that almost every man should be a man of business, or engage in some sort of trade. The province of Holland seems to be approaching near to this state. It is there unfashionable not to be a man of business. Necessity makes it usual for almost every man to be so, and custom every where regulates fashion.[78]

England would soon reach this plateau and then, like Holland, became stuck in one form of the high-level equilibrium trap.[79]

A second mechanism, which partly stemmed from the first, was the law of diminishing marginal returns, particularly in agriculture. This was more famously and explicitly enunciated by Malthus and Ricardo, but it was also obvious to Smith. Put simply, new land produces a good harvest, but as demands continue it produces less, and the use of marginal lands, or the application of extra labour brings decreasing returns. The principle of the division of labour had temporarily overcome part of this problem, but the marginal returns on the division of labour also began to reach a limit. Mankind was trapped on a treadmill which required more and more effort for less and less returns. As Wrigley notes, the restraints which seemed to be 'permanent and ineradicable' in Smith's world were that land was the source of all wealth, and that energy was limited to what could be obtained directly from the sun, wind and water.[80]

The third law that trapped mankind was that of population. In a direct anticipation of Malthus, Smith explained how the history of agrarian societies showed that 'men, like all other animals, naturally multiply in proportion to the means of their subsistence'.[81] Thus whenever the wealth of a nation increased, and in particular if this wealth was shared by the mass of the population through higher real wages, the population would increase to absorb the increase. This is the point which Smith stresses in both volumes of his book. In the first he notes that, as he puts it in the heading, 'High wages increase population'. 'The liberal reward of labour, therefore, as it is the effect of increasing wealth, so it is the cause of increasing population. To complain of it, is to lament over the necessary effect and cause of the greatest public prosperity.'[82] Or in a more expanded way,

> The liberal reward of labour, by enabling them to provide better for their children, and consequently to bring up a greater number, naturally tends to widen and extend those limits. It deserves to be remarked too, that it necessarily does this as nearly as possible in the proportion which the demand for labour requires. If this demand is continually increasing, the reward of labour must necessarily encourage in such a manner the marriage and multiplication of labourers, as may enable them to supply that continually increasing demand by a continually increasing population.[83]

The law of supply and demand works with population as with anything else. Thus 'the demand for men, like that for any other commodity, necessarily regulates the production of men'.[84]

The danger was even greater because poverty in itself did not necessarily limit population growth. 'Poverty, though it no doubt discourages, does not always prevent marriage. It seems even to be favourable to generation. A half-starved Highland woman frequently bears more than twenty children, while a pampered fine lady is often incapable of bearing any...' What poverty did do, he thought, was to kill off large numbers of infants: 'in the Highlands of Scotland it is not uncommon for a mother who has borne twenty children not to have two alive.'[85] Thus, if the standard of living and medical care of the poor increased markedly, the problem of population growth would be even greater. Mankind was caught in the Malthusian trap. Every short-term gain would lead to a larger problem in the future.

Thus, placing ourselves in Adam Smith's world as he sat beside the Firth of Forth slowly compiling the *Wealth of Nations* in the years before 1776, we can see how he must have felt clearly both grounds for measured short-term optimism and long-term pessimism. The 'natural' path to increased opulence was there to be taken if the mainly political obstacles could be removed. Everyone could, in theory, reach the level of the Dutch. But then people were trapped on a high plateau. Although they were not so vulnerable to external destruction and predation, there were reasons for suspecting that having reached the plateau, the only path was downwards. Growing population, the monopolistic tendencies of greedy merchants or even farmers, the ambitions of the State, the ambitions of the Church, any or all of these could shatter the precarious balance of forces.

This is ultimately Smith's message. Although there was a 'natural tendency' for the selfish and competitive drives of human beings to lead to the growth of wealth if appropriate conditions were provided, continuous, unlimited growth was impossible. The growth in the past had been the unintended consequence of a set of accidents – outcomes of conflicts and oppositions which had against all the odds led to gradual growth. Many had strayed from the path – eastern Europe, later much of southern Europe, China and India. Even France was in doubt, and Germany is hardly mentioned. Only on an outlying tip of north-west Europe, and in the New World, was conspicuous growth still occurring. It is not surprising, therefore, that Adam Smith, like Malthus and Ricardo 'unanimously and explicitly denied the possibility of the change now regarded as its [industrial revolution] most important single feature, and perhaps as its great redeeming feature – the substantial and largely continuous rise in the standard of living that it has occasioned'.[86] Mankind was trapped at a high-level equilibrium.

It was also trapped in another way. A number of commentators have pointed out that Smith anticipates Marx concerning some of the disastrous side-effects of the new industrial capitalism which he saw emerging around him.[87] He had observed that alongside the growing wealth, even in the richest parts of one of the richest countries in the world, England, there was increasing misery, although part of this was self-inflicted.

> Accordingly we find that in the commercial parts of England, the tradesmen are for the most part in this despicable condition; their work through half the week is sufficient to maintain them, and through want of education they have no amusement for the other but riot and debauchery. So it may very justly be said that the people who clothe the whole world are in rags themselves.[88]

This was no accident, for it rose from the very essence of the new division of labour which was the motor of change. He had noticed that 'It is remarkable that in every commercial nation the low people are exceedingly stupid. The Dutch vulgar are eminently so, and the English are more so than the Scotch. The rule is general, in towns they are not so intelligent as in the country, nor in a rich country as in a poor one.'[89] This was not because of some innate inferiority, but because of the crippling effects of a life making pinheads. Partly there was the sheer pettiness and boredom of the activity.

> Where the division of labour is brought to perfection, every man has only a simple operation to perform. To this his whole attention is confined, and few ideas pass in his mind but what have an immediate connection with it. When the mind is employed about a variety of objects it is some how expanded and enlarged, and on this account a country artist is generally acknowledged to have a range of thoughts much above a city one.[90]

Partly it was because education was brushed aside in the rush to use the labour of children.

> Another inconvenience attending commerce is that education is greatly neglected. In rich and commercial nations the division of labour, having reduced all trades to very simple operations, affords an opportunity of employing children very young. In this country indeed, where the division of labour is not far advanced, even the meanest porter can read and write, because the price of education is cheap, and a parent can employ his child no other way at 6 or 7 years of age.[91]

He concluded, 'These are the disadvantages of a commercial spirit. The minds of men are contracted and rendered incapable of elevation, education is despised or at least neglected, and heroic spirit is almost utterly extinguished. To remedy these defects would be an object worthy of serious attention.'[92] It was a serious attention which Smith himself, unfortunately, was unable to provide. Indeed, since he did not fully appreciate the liberating effects of machinery, it was difficult, if not impossible, for him to see a way round these difficulties. Thus both at the national and individual level, the 'wealth of nations' was tinged with failure. Mankind had not escaped from the treadmill of existence, even if the present condition in a few favoured nations was perhaps better than it had been since the descent from the 'original affluent society' of hunter-gathering.

*

In terms of Smith's solution to the riddle, he confirmed certain parts of the answer already suggested by Montesquieu. He noted the normal tendency to stasis, the beneficial effects of commerce, the difficulties caused by the size, homogeneity and rice cultivation in China, the dangers of conquest and war, the importance of English Common Law, the importance of a reasonable taxation system and secure investment opportunities, and the advantage of being an island.

New areas of the puzzle now filled in included the discussion of the rise and effect of towns, of the middle class, the night watchman state and church. He put forward the theory that liberty emerges when sects fall out with each other and adds to this a description of the effects of commercial wealth on power. His account of the mechanism for the escape from violence through the growth of opulence is extremely suggestive. And his reworking of the theory of the division of labour provides some dynamic for the change. His account of English development adds detail to Montesquieu and reinforces the importance of islandhood. Smith warned of the dangers of all monopolies of power, even those of producers and exchangers, he noted the role of the judiciary in safeguarding economic wellbeing and he noted the importance of the unification of England and Scotland. He even anticipated some of the negative effects of the division of labour and a commercial mentality on the morals and wellbeing of future generations.

Yet even when we add his formidable contribution to that of Montesquieu, the riddle is still partly unresolved. The fact that Smith was pessimistic about the future shows that he did not solve it. Part of the answer lay in the development of science and industrial technology which he only glimpsed. He was on the whole unaware of the power of the scientific revolution, that is the growth of new knowledge through the use of the experimental method, which provided the basis for the new manufacture of artifacts through the industrial process. Nor did he fully realize that the rapid growth of England was dependent on its position as part of a European network of knowledge. We might say that after the contributions of Montesquieu and Smith the solution was half complete. Like many operations, the relatively easier parts are done first. To fill in the last parts is the most difficult and it is indeed fortunate that in Tocqueville we find a thinker fit for the task of adding some of the important pieces.

III Equality

9
Alexis de Tocqueville's Life and Vision

Alexis-Charles-Henri de Tocqueville was born in Paris on 29 July 1805. He was the son of Count Herve (landed proprietor and prefect) and Louise de Tocqueville, and the great-grandson of Lamoignon de Malesherbes, an eighteenth-century statesman of renown. Tocqueville was of noble descent on both his father's and mother's side and the family now had its main estates in Normandy. His parents had suffered badly during the French Revolution: they had been imprisoned, and came within a few days of being guillotined.

Tocqueville was tutored by the Abbé Lesueur, an important moral and intellectual influence upon him and largely brought up by his father. He then attended the lycée at Metz until 1823. From 1823 to 1827 he studied law in Paris. In 1826–7 he travelled in Italy and Sicily with his brother. He served as a *juge-auditeur* (magistrate) at the Versailles Tribunal from 1827 to 1831. During this period he attended Guizot's lectures on the history of Europe and philosophy of history and became engaged to be married to an English lady, Mary Mottley.

From May 1831 to February 1832 Tocqueville visited America with Gustave de Beaumont. They travelled as far north as Quebec and as far south as New Orleans. In 1833 he went for five weeks to England, and from September 1833 he spent 12 months writing the first volume of *Democracy in America*, which was published in 1835. He also made a second, longer trip to England from May to September 1835. In October 1836 he married Mary Mottley and travelled to Switzerland.

In 1837 Tocqueville failed to get elected to the Chamber of Deputies, but he did achieve this in 1839. During these years he had been writing the second volume of *Democracy in America*, which was published in 1840. In 1841 he was elected a member of the French Academy and travelled with Beaumont to Algeria. He was elected to the General Council of La Manche in 1842 and later became president. From 1841–3 he worked on a study of India. In 1844–5 he became involved in a progressive newspaper, *Le Commerce*, which advocated various liberal programmes. In 1846 he made a second trip to Algeria with his wife.

In 1848 Tocqueville made a speech to the Chamber warning of the coming Revolution, and in that year was elected to the Constituent Assembly and was involved in writing a new constitution. In 1849 he was elected to the new Legislative Assembly and was briefly minister of foreign affairs. In 1850–1 he wrote *Recollections*, an account of the period 1848–51. In December 1851 he and other members of the Assembly opposed a coup, and he was arrested and held for one day. In 1853 Tocqueville started to study in the archives at Tours as a preparation for his work on the *Ancien Régime*. In 1854 he travelled to Germany to study feudalism and social structure. In 1856 he published the *Ancien Régime*. In 1857 he visited England again and was greeted with high acclaim. On 16 April 1858 he died at Cannes, aged 53.

*

What strikes one most forcefully about Tocqueville's life is that the central motif behind his work was a set of contradictions, which he was always seeking to resolve in his writing.[1] 'I passionately love liberty, legality, the respect for the law, but not democracy; that is the deepest of my feelings.'[2] In a discarded note a different formulation was 'Mon Instinct, Mes Opinions.' 'I have an intellectual taste for democratic institutions, but I am an aristocrat by instinct, that is I fear and scorn the mob' (*la foule*).[3] He wrote to a friend in 1835: 'I love liberty by taste, equality by instinct and reason. These two passions, which so many pretend to have, I am convinced that I really feel in myself, and that I am prepared to make great sacrifices for them.'[4] The clash between his mind and his heart was caught by Sainte-Beuve when he wrote that Tocqueville's whole doctrine had been 'a marriage of reason and necessity, not at all of inclination'.[5] As Pierson writes, 'Wrestling with contrary impulses, his spirit torn by opposing loyalties, his career was to be one long, never-ending struggle to reconcile the powerful forces clashing for mastery within him. In the end, it was only as a crier in the wilderness, only as the solemn, foreboding prophet of equality that he was to achieve some measure of spiritual peace.'[6]

This clash between the aristocratic and democratic sides of his nature meant that although he had always refused to use the title of comte, he remained attached to his aristocratic family line. In 1858 just before he died he wrote to his wife:

> We will not be replaced, as I often tell myself sadly... We are part... of a world that is passing. An old family, in an old house that belonged to its forefathers, still enclosed and protected by the traditional respect and by memories dear to it and to the surrounding population – these are the remains of a society that is falling into dust and that will soon have left no trace. Happy are those who can tie together in their thoughts the past, the present, and the future! No Frenchman of our time has this happiness and already few can even understand it.[7]

He summarized the reasons for his own ambivalence in a letter in 1837.

> All forms of government are in my eyes only more or less perfect ways of satisfying this holy and legitimate passion of man. They alternately give me democratic or aristocratic prejudices; I perhaps would have had one set of prejudices or the other, if I had been born in another century and in another country. But the chance of birth has made me very comfortable defending both. I came into the world at the end of a long Revolution, which, after having destroyed the old state, had created nothing durable. Aristocracy was already dead when I started life and democracy did not yet exist, so my instinct could lead me blindly neither toward one nor toward the other. I was living in a country that for forty years had tried a little of everything without settling definitely on anything; therefore I was not susceptible to political illusions. Belonging to the old aristocracy of my homeland, I had neither hatred nor natural jealousy against the aristocracy, and that aristocracy being destroyed, I did not have any natural love for it either, since one only attaches oneself strongly to what is living. I was near enough to it to know it well, far enough away to judge it without passion. I would say as much about the democratic element. No family memory, no personal interest gave me a natural and necessary bent toward

democracy. But for my part I had not received any injury from it; I had no particular motive for either loving or hating it, independent of those that my reason furnished me. In a word, I was so thoroughly in equilibrium between the past and the future that I felt naturally and instinctively attracted toward neither the one nor the other, and I did not need to make great efforts to cast calm glances on both sides.[8]

It was this placing halfway between which allowed him to see so clearly. It led him to advocate a middle road which was both revolutionary and conservative, monarchist and republican, centralizing and decentralizing. He gave a summary of this creed in a letter of 1836.

I do not think that in France there is a man who is less revolutionary than I, nor one who has a more profound hatred for what is called the revolutionary spirit (a spirit which, parenthetically, is very easily combined with the love of an absolute government). What am I then? And what do I want? Let us distinguish, in order to understand each other better, between the end and the means. What is the end? What I want is not a republic, but a hereditary monarchy. I would even prefer it to be legitimate rather than elected like the one we have, because it would be stronger, especially externally. What I want is a central government energetic in its own sphere of action ... But I wish that this central power had a clearly delineated sphere, that it were involved with what is a necessary part of its functions and not with everything in general, and that it were forever subordinated, in its tendency, to public opinion and to the legislative power that represents this public opinion.[9]

He was aware of the difficulty of achieving this balance between contrary pressures, yet believed, as shown in the same letter, that 'all these things are compatible', and 'that there will never be order, and tranquillity except when they are successfully combined'.[10]

As to whether they would be combined, and that he and France and the world would reach tranquillity, he was not sure. Just as his personality was a mixture of hope and despair, so his writings are an exact blend of pessimism and optimism about the future, as well as the past and the present. Towards the end of the second volume of *Democracy in America* he wrote that 'I find that good things and evil in the world are fairly evenly distributed'.[11] He noted that 'Men tend to live longer, and their property is more secure. Life is not very glamorous, but extremely comfortable and peaceful.'[12] A middling condition had been attained. 'Almost all extremes are softened and blunted. Almost all salient characteristics are obliterated to make room for something average, less high and less low, less brilliant and less dim, than what the world had before.'[13]

Yet he was also full of fear and regret. 'When I survey this countless multitude of beings, shaped in each other's likeness, among whom nothing stands out or falls unduly low, the sight of such universal uniformity saddens and chills me, and I am tempted to regret that state of society which has ceased to be.'[14] But the worst might never happen. 'I am full of fears and of hopes. I see great dangers which may be warded off and mighty evils which may be avoided or kept in check; and I am ever increasingly confirmed in my belief that for democratic nations to be virtuous and prosperous, it is enough if they will to be so.'[15] Laski suggests that his later work, the *Ancien Régime*, is even more uncertain and pessimistic.[16] Certainly Tocqueville felt exactly balanced between the two emotions of hope and despair, and this was a feeling which he seems to have had during much of his life.

This then was the man who stands in the tradition of Montesquieu and Smith as one of the deepest thinkers about the riddle of the modern world. At every level his experiences placed him in a position to stand outside the great turmoils of the time. Yet he was close enough to them to be able to see their inner causes. As he put it, writing specifically of the French Revolution,

> It would seem that the time for examination and judgment on it has arrived. We are placed to-day at that precise point, from which this great subject can be best perceived and judged. We are far enough from the Revolution not to feel violently the passions which disturbed the view of those who made it. On the other hand we are near enough to be able to enter into and to understand the spirit which produced it. Very soon it will be difficult to do so. For great successful revolutions, by effecting the disappearance of the causes which brought them about, by their very success, become themselves incomprehensible.[17]

In order to analyse and try to understand the puzzles and confusions that faced him as the industrial and political revolutions took their hold, he needed other weapons beyond deep sensitivity and a brilliant mind. He needed a theoretical system and wide experience of a changing world.

*

The essence of Tocqueville's method, as it was of Montesquieu's, was to try to penetrate to 'the Spirit of the Laws', that is to say the principles which generated the system.[18] And again, like Montesquieu, this spirit was not composed of *things*, but relations between things – between liberty and equality, individual and group, centre and periphery. What he sought to do was to practice a kind of mental cartography, to discern the plan or map behind a civilization – how it was laid out. He commended the 'sagacity which penetrates through the passions of the time and of the country, down to the general character of an epoch, and to its place in human progress.'[19]

Sometimes the pattern was simple and symmetrical, as in a new country like America which is relatively easy to understand. 'The man whom you left in the streets of New York you find again in the solitude of the Far West; the same dress, the same tone of mind, the same language, the same habits, the same amusements.'[20] There is less difference over the thousands of miles in America than there is between the tens of miles between different regions of France. Thus, 'In America, more even than in Europe, there is but one society, whether rich or poor, high or low, commercial or agricultural; it is everywhere composed of the same elements. It has all been raised or reduced to the same level of civilization.'[21] The principle of America is equality, and this generates everything.

> In America all laws originate more or less from the same idea. The whole of society, so to say, is based on just one fact: everything follows from one underlying principle. One could compare America to a great forest cut through by a large number of roads which all end in the same place. Once you have found the central point, you can see the whole plan in one glance. But in England the roads cross, and you have to follow along each one of them to get a clear idea of the whole.[22]

England is an old country, where there are contradictions and inconsistencies, and the winding tracks of a thousand years of history. William the Conqueror had set up a consistent system of government: 'the system made a more coherent whole than in any other country, because one head had thought out all the machinery and so each wheel fitted better.'[23] Yet over time it had evolved and twisted into new shapes. In America, with its sparse population and short history this had not happened. It lacked the contradictions of class and the overgrowths of one system superimposed on another that one found in European countries. When he arrived in England he expressed the contrast thus.

> So far this country seems to me, still, to be one vast chaos. This is certainly a different sort of difficulty to overcome than in the study of America. Here, there is not that single principle which tranquilly awaits the working out of its consequences, but instead lines that cross one another in every direction, a labyrinth in which we are utterly lost.[24]

Much of Tocqueville's brilliance arises out of his explicitly comparative method. He wrote: 'no one, who has studied and considered France alone, will ever, I venture to say, understand the French Revolution'.[25] At more length he summarized his method as follows.

> In my work on America. . . . Though I seldom mentioned France, I did not write a page without thinking of her, and placing her as it were before me. And what I especially tried to draw out, and to explain in the United States, was not the whole condition of that foreign society, but the points in which it differs from our own, or resembles us. It is always by noticing likenesses or contrasts that I succeeded in giving an interesting and accurate description . . .[26]

Again and again on his American tour he stressed this necessity. 'In this examination, one great obstacle arrests me. Each fact is without particular physiognomy for me, and without great significance because I can make no comparisons. Nothing would be more useful for judging America well than to know France.'[27] Thus he testifies to the fact that France was always in his mind, night and day, as he observed America. 'In the midst of all the theories with which I am amusing my imagination here, the memory of France is becoming like a worm that is consuming me. It manages to surprise me by day in the midst of our work, by night when I wake up.'[28] In fact, by making a three-way triangulation of France, England and America he was able to develop an especially powerful version of the comparative method.[29]

The problem was how one was to grasp the whole of a civilization for comparative purposes. Tocqueville stressed the difficulty on a number of occasions. 'Every foreign nation has a peculiar physiognomy, seen at the first glance and easily described. When afterwards you try to penetrate deeper, you are met by real and unexpected difficulties; you advance with a slowness that drives you to despair, and the farther you go the more you doubt.'[30] It was important to grasp the first impressions of another country, 'For he had remarked that the first impression gives itself utterance almost always in an original shape, which, once lost, is not recovered.'[31] Yet this first impression was only that. 'It would take a very fatuous philosopher to imagine that he could understand England in six months. A year has ever seemed to me too short a time for a proper appreciation of the United States, and it is infinitely easier to form clear ideas and precise conceptions about America than about Great Britain.'[32] Indeed, at times, he thought the task was impossible. 'You are right when you say that a foreigner cannot understand the peculiarities of the English character. It is the case with almost all countries.'[33] Yet one should still attempt to penetrate this otherness, even if it meant, in true anthropological fashion, a kind of willing suspension of disbelief or almost surrendering one's identity. 'I do not know how national character is formed, but I do know, that when once formed, it draws such broad distinctions between nations, that to discover what is passing in the minds of foreigners, one must give up one's own nationality, almost one's identity.'[34]

His basic aim was to see how the separate parts of a social system work and are connected together into a general, functioning, integrated whole. He may have received much of this vision from Montesquieu, whom we have seen also espoused such an approach.[35] He was also strongly influenced by Guizot. For example in his notes on a lecture by Guizot on 18 July 1829, Tocqueville wrote:

> the history of civilization ... should and does try to embrace everything simultaneously. Man is to be examined in all aspects of his social existence. History must follow the course of his intellectual development in his deeds, his customs, his opinions, his laws, and the monuments of his intelligence ... In a word,

it is the whole of man during a given period that must be portrayed...³⁶

This involved both general theory and an attention to the smallest details. The use of the microscope was as important as that of the telescope. Thus he wrote during his last visit to England in 1857:

Besides, there is not a single one of my theoretical ideas on the practice of political liberty and on what allows it to function among men that does not seem to me fully justified once again by everything I have been seeing before me. The more I have delved into the detail of the way in which public affairs are conducted, the more these truths seem to me to be demonstrated: for it is the manner in which the smallest of affairs are managed that leads to a comprehension of what is happening in the great ones. If one were to limit oneself to studying the English political world from above, one would never understand anything about it.³⁷

Yet while delving into the minutiae, it was always necessary to connect each of these details into something larger. 'Is it enough to see things separately, or should we discover the hidden link connecting them?'³⁸ His answer is clear in his writings. 'He always attempted to convert specific observations into the broadest generalities that the fact at hand could be made to bear...'³⁹ When he did this and his readers failed to see the links he had made he became upset. He wrote to Stoffels, having explained the purpose of the first volume of *Democracy*, 'There is the mother-idea of the work, the idea which links all the others in a single web, and which you should have perceived more clearly than you did.'⁴⁰ The web metaphor hints at his aims. Even while exploring a particular thread or track, be aware of how it fits into the whole. He never became too involved in either thread or web, but kept a balance between them.

*

As a disciple of Montesquieu, Tocqueville was an heir to a mixed inheritance but one which put quite a heavy emphasis on geographical determinism. Thus when he went to America he expected this vast new world with its dramatic geography and climate and sparse population to show the predominant influence of the ecology. In fact, what he found shocked him. 'By a strange inversion of the ordinary order of things, it is nature that changes, while man is unchanging.'[41] One example was the contrast between the French and the English parts of Canada. Despite a similar ecology, the two groups of settlers were entirely different. He found the extreme case when he travelled into the wildest part: 'The inhabitants of this little oasis belong to two nations which for more than a century have occupied the same country and obeyed the same laws. Yet they have nothing in common. They still are as distinctly English and French as if they lived on the banks of the Seine and the Thames.'[42] He saw it clearly at a higher level in the difference between the English-settled world of North America, and the Spanish and Portuguese parts of South America.[43]

His next theory concerning the causes of things followed another strand in Montesquieu's thought, that is to say 'The Spirit of the Laws'. As Lerner writes, 'He learned relatively early to regard legal custom, statute, and code as keys for unlocking the inner meaning of social structure and national character. On this score the influence of Montesquieu and his *L'Esprit des Lois* on his thinking must be considered a capital one.'[44] But even this was not enough. Tocqueville began to realize that 'there must be some other reason, apart from geography and laws, which makes it possible for democracy to rule the United States.'[45] This 'other reason' was what anthropologists term 'culture'. 'The importance of mores is a universal truth to which study and experience continually bring us back. I find it occupies the central position in my thoughts; and all my ideas come back to it in the end.'[46] He had found the key. 'It is their mores, then, that make the Americans of the United States, alone among Americans, capable of maintaining the rule of democracy; and it is mores again that make the various Anglo-American democracies more or less orderly and prosperous.'[47]

How could one explain these *mores*? They did not just suddenly appear, and they varied so surprisingly between cultures. Here he developed one of his most important ideas. Drescher describes how

> It was also in connection with the analysis of American self-government that Tocqueville and Beaumont hit upon a primary organizational concept for their later works – the idea of the 'point de départ', or point of departure. Methodologically, an inductive discovery of the basic tendencies or fundamental social

fact of the present led to a historical search for the original act or circumstances from which the present could be seen to have unfolded.⁴⁸

He then points out that 'From the *Démocratie* to the *Ancien Régime*, unless Tocqueville could discover a social context with objectively discernible characteristics from which all subsequent developments could be logically explained, he did not feel that he had successfully encompassed the problem.'⁴⁹

Drescher quotes Tocqueville to the effect that 'One can't help being astonished at the influence, for good or evil, of the point of departure on the destiny of peoples'.⁵⁰ This can be parallelled by many similar observations in his works. In his notebooks of the American trip he wrote, when listing the causes of what he saw before him, '1st. *Their origin*: Excellent point of departure. Intimate mixture of the spirit of religion and liberty. Cold and rationalist race.'⁵¹ In the first volume of *Democracy* he stresses this approach. Nations, like people, are deeply influenced by their birth and formative years. 'People always bear some marks of their origin. Circumstances of birth and growth affect all the rest of their careers.' 'Something analogous happens with nations.'⁵² Thus, in general, he believed of nations, as of individuals, that 'If we could go right back to the elements of societies and examine the very first records of their histories, I have no doubt that we should there find the first cause of their prejudices, habits, dominating passions, and all that comes to be called the national character.'⁵³

This was particularly obvious in the case of a 'new' nation like America. 'When, after careful study of the history of America, we turn with equal care to the political and social state there, we find ourselves deeply convinced of this truth, that there is not an opinion, custom, or law, nor, one might add, an event, which the point of departure will not easily explain.'⁵⁴ Putting it in an extreme and aphoristic form, he came to believe that 'When I consider all that has resulted from this first fact, I think I can see the whole destiny of America contained in the first Puritan who landed on those shores, as that of the whole human race in the first man.'⁵⁵

It was this insight that makes his later reflections on the nations of 'old' Europe so rich. He realized how important it was to trace the history of present structures back into the past. Particularly in the *Ancien Régime* he gave a brilliant exposition of the way in which certain ideas spread out from a particular 'point of origin' until they came to influence the whole of a civilization. In a footnote to that work he explained how

> Every institution that has long been dominant, after establishing itself in its natural sphere, extends itself, and ends by exercising a large influence over those branches of legislation which it does not govern. The feudal system, though essentially political, had transformed the civil law, and greatly modified the condition of persons and property in all the relations of private life.'[56]

This shows that the 'point of origin' was not a static concept. He saw a set of ideas changing and branching. It is an organic metaphor which could be interpreted as a partial anticipation of that evolutionary paradigm which was already widespread in the minds of Wallace, Darwin, Robert Chambers, Herbert Spencer and others, even if the *Origin of Species* was still three years from publication.

*

Tocqueville was well aware of the need for precision in the use of key terms. For instance, he wrote 'I would like to take apart the word *centralization*, which, by virtue of its vague immensity, wearies the mind without leading it to anything.'[57] Yet he seems to have left his most important words, democracy and equality, deliberately ambiguous. Part of the difficulty was pointed out by J.S. Mill in the review of Volume I of *Democracy* in 1835:

> M. de Tocqueville then has, at least apparently, confounded the effects of Democracy with the effects of Civilization. He has bound up in one abstract idea the whole of the tendencies of modern commercial society, and given them one name – Democracy; thereby letting it be supposed that he ascribes to equality of conditions, several of the effects naturally arising from the mere progress of national prosperity, in the form in which that progress manifests itself in modern times.[58]

It is clear that Tocqueville himself realized that he had failed to define or distinguish his two key terms. Drescher points out that 'In the notes for the *Démocratie* of 1840 Tocqueville had considered drawing a distinction between "democratie" and "égalité": "When I understand [the new society] in the political sense, I say

"Démocratie". When I want to speak of the effects of equality, I say "égalite."'[59] Yet Drescher also points out that 'This clarification, whether because it would have aesthetically weakened the impact of the term, or for some other reason, remained buried in his papers and his book went to press with "equality" and "democracy" used interchangeably.'[60] Others have also noted the ambiguities. Pierson asks 'how he ever allowed himself to use "démocratie" in seven or eight different senses is still something of a mystery. It was his key word.'[61] It appears that Tocqueville found it logically unsatisfactory to split the two. Indeed his skill lay in connecting, in holding pairs in tension. Here he fused two separate meanings into one and his work would have been clearer but less insightful if he had subsequently split them again. As he might have put it, tranquillity and peace of mind might have been gained – but at the price of logical interconnections.

The other main criticism of his approach lies in the assertion that, particularly in his later work, as he moved further away from the 'facts' of America, he came to rely too much on the deductive method; in other words he worked out the theories first and fitted the facts to them, rather than keeping a blend between them. Two of his wisest contemporaries alluded to such a charge. Lerner writes that even when he went to America 'Saint-Beuve's famous quip about the young Tocqueville, that "he began to think before having learned anything," has a light sting of truth in it. There is little question that he had a whole trunkful of ideas stored away in his mind, the result of his reading of the political classics, his work as a magistrate, his observation of men and nations.'[62] Royer-Collard tried to explain why the 'prodigious effort of meditation and patience' of the second volume of America had caused misunderstanding, writing that Tocqueville was constructing ideal types, a procedure with which people were not familiar. 'There is not one chapter that could not be different in certain respects from the way you have done it. That, of course, is because of your intention. You set out to imagine, to invent rather than to describe, and invention, within certain limits, is arbitrary.'[63]

Tocqueville himself felt hurt by these charges, for he believed that 'I have never knowingly moulded facts to ideas instead of ideas to facts.'[64] He perhaps took comfort from the views of the greatest nineteenth-century expert on logical methods in the social and physical sciences, J.S. Mill. Mill pointed out that, on the surface, there were indeed grounds for doubt, 'It is perhaps the greatest defect of M. de Tocqueville's book, that from the scarcity of examples, his propositions, even when derived from observation, have the air of mere abstract speculations.'[65] Nevertheless he asserted that

> The value of his work is less in the conclusions, than in the mode of arriving at them. He has applied to the greatest question in the art and science of government, those principles and methods of philosophizing to which mankind are indebted for all the advances made by modern times in the other branches of the study of nature. It is not risking too much to affirm of these volumes, that they contain the first analytical inquiry into the influence of democracy.[66]

He believed that Tocqueville had blended the two approaches. 'His method is, as that of a philosopher on such a subject must be – a combination of deduction with induction: his evidences are laws of human nature, on the one hand; the example of America and France, and other modern nations, so far as applicable, on the other.'[67]

Mill's summation places Tocqueville as the man who combined the deductive and the inductive methods.

> His conclusions never rest on either species of evidence alone; whatever he classes as an effect of Democracy, he has both ascertained to exist in those countries in which the state of society is democratic, and has also succeeded in connecting with Democracy by deductions *a priori*, showing that such would naturally be its influences upon beings constituted as mankind are, and placed in a world such as we know ours to be. If this be not the true Baconian and Newtonian method applied to society and government...[68]

Mill concluded his assessment with an affirmation of Tocqueville's genius: 'though we would soften the colours of the picture, we would not alter them; M. de Tocqueville's is, in our eyes, the true view of the position in which mankind now stand...'[69]

10
'America' as a Thought Experiment

Tocqueville described his visit to America in 1833 as a second discovery of that world.[1] He spent 286 days in the New World and then wrote the two volumes which form the greatest anthropological essay on a civilization that we possess. Yet before briefly outlining his findings and hypotheses, we need to be clear about two things. The first is that the two volumes of his *Democracy*, as many Tocqueville scholars have observed, are really very different works. He should really have called them something like 'America', and 'Democracy' to prevent confusion.[2] In the following account, for brevity and coherence, I will treat them together, though not only time but shifting intentions made them feel very different.

The major difference is alleged to be that the first book was really about America, and the second just used America as a way of talking about equality (democracy). Certainly there is a shift. But it is also important to grasp that even the first volume was really a way of experimenting with ideas that Tocqueville had partially worked out before his visit. The point is well made by Pierson. 'It will not escape the student that Tocqueville had just reversed the sequence of his perceptions. For literary purposes he implied that he had discovered his great natural law of modern societies in America. Actually, this idea had been the product of his youthful experiences at home.'[3]

In fact Tocqueville was fairly open about this. Near the start of the first volume he wrote 'I admit that I saw in America more than America; it was the shape of democracy itself which I sought, its inclinations, character, prejudices, and passions; I wanted to understand it so as at least to know what we have to fear or hope therefrom.'[4] In 1834 between the visit and the publication, he wrote to a friend.

> Some will find that at bottom I do not like democracy and that I am severe toward it; others will think that I favour its development imprudently... but this is my response: nearly ten years ago I was already thinking about part of the things I have just now set forth. I was in America only to become clear on this point. The penitentiary system was a pretext: I took it as a passport that would let me enter thoroughly into the United States.'[5]

Or as he put it succinctly commenting on the first volume, in a letter to J.S. Mill, 'America was only my framework; democracy was my subject.'[6]

His reasons for choosing America seem to have been threefold. Firstly, America presented a simple, clear, field for the investigation of the questions that interested him. 'The special reason that has put the Americans in a state to be understood, is that they have been able to build their social edifice from a clean start.'[7] Secondly, with his fear of the emptiness of the abstract, he felt that his message would have more power if written as a kind of narrative. He himself recognized the rhetorical need to make the abstract concrete, as had Montesquieu and Smith, and he wondered explicitly at one point: 'Here I want to illustrate how the government can do things which no power before it had done. But perhaps this idea might be introduced in narrative form...'[8] In many ways all of his work is a semi-narrative, a journey or exploration, the outward form is America, England or France, the inner thought is mankind and the riddle of modern civilization.

The third aim was to use America as a guide. 'So I did not study America just to satisfy curiosity, however legitimate; I sought there lessons from which we might profit.'[9] America should not, of course, be directly imitated. 'The new society in which we are does not at all resemble our European societies. It has no prototype anywhere. It has also some primary conditions of existence that no other possesses, which makes it dangerous for any other society to imitate it...'[10] Yet one could learn from it, and particularly from its mistakes. To make a 'mistake' in an old civilization like Europe was usually disastrous. But the energy, youth and flexibility of America meant that 'the great privilege of the Americans is to be able to make retrievable mistakes'.[11] Thus America was a thought experiment in more than one sense. It was a place for Tocqueville to test ideas he had been developing since he was nineteen, but it was also, in itself, a civilization which was making mistakes and retrieving itself

– trying out new things and hence showing old Europe what it should and should not do.

*

Tocqueville partly chose America because it appeared to be young, sparsely populated, relatively 'simple'. He came to think of it as laid out on a kind of grid. If one could understand the straight roads, then one could understand everything. Once he arrived there he found another advantage, which was its homogeneity. Leaving on one side French Canada, he found that there was a surprising similarity at a deep level over the whole continent – despite dramatic geographical differences. This contrasted enormously with his experience of Europe. 'I doubt whether there is any nation in Europe, however small, whose different parts are not less homogeneous than those of the United States with an area half the size of Europe.'[12] More specifically, and recalling his own experience of the northwest tip of France, he wrote: 'From the state of Maine to that of Georgia is a distance of some thousand miles, but the difference in civilization between Maine and Georgia is less than that between Normandy and Brittany.'[13]

The combination of newness and homogeneity made it possible to comprehend and even to give it a name, 'America'. Yet it was no easy task to understand this new civilization. There were at first arrival a welter of impressions, often confusing, which had to be sorted out. 'You understand that I cannot yet have a fully developed opinion of this people. At first sight, it presents, like all others, a mixture of vices and virtues that is rather difficult to classify and that does not form a single picture.'[14] He sensed that there was a surface and a deeper structure to be understood. As he later put it, 'The vices and weaknesses of democratic government are easy to see; they can be proved by obvious facts, whereas its salutary influence is exercised in an imperceptible and almost secret way.'[15] Yet the difficulty was not merely one at the level of confusion, or of surface and base. The real difficulty, as Tocqueville realized, was that America was built on contradictions. It was the outcome of logically incompatible elements. Its fascination came from the fact that it was a *new* mix of forces, forming a restless, seething set of combinations.

Tocqueville uses an image of a pool in a rushing stream where contrary flows meet and swirl when he is trying to capture one of the major contradictions. 'When one examines what is happening in the United States closely, one soon discovers two contrary tendencies; they are like two currents flowing in the same bed in opposite directions.'[16] At other times he described more than two contrary flows. For instance he talked of 'the great American fight between the provinces and the central power, between the spirit of independence and democracy, and the spirit of hierarchy and subordination.'[17] He noted the agitation. 'This constant strife between the desires inspired by equality and the means it supplies to satisfy them harasses and wearies the mind.'[18] But paradoxically it also led to order. 'One may say that it is the very vehemence of their desires that makes the Americans so methodical. It agitates their minds but disciplines their lives.'[19] If there was a central feature to the New World that was to be seen in America it was its turbulence and restlessness, its absence of tranquillity.

Tocqueville described this turbulence caused by the conflicts of desire and reason, centre and periphery, equality and individualism, in a number of brilliant passages. 'No sooner do you set foot on American soil than you find yourself in a sort of tumult; a confused clamour rises on every side, and a thousand voices are heard at once, each expressing some social requirements. All around you everything is on the move.'[20] The contrast was re-emphasized when he returned to France.

> When one passes from a free country into another which is not so, the contrast is very striking: there, all is activity and bustle; here all seems calm and immobile. In the former, betterment and progress are the questions of the day; in the latter, one might suppose that society, having acquired every blessing, longs for nothing but repose in which to enjoy them.[21]

The restless, swiftly changing cascade is what struck him forcefully.

> Restlessness of character seems to me to be one of the distinctive traits of this people. The American is devoured by the longing to make his fortune; it is the unique passion of his life; he has no memory that attaches him to one place more than another, no inveterate habits, no spirit of routine; he is the daily witness of the swiftest changes of fortune.[22]

Again he invoked the metaphor of a rushing stream with its crosscutting currents and turbulent cataracts. 'Often born under another sky, placed in the middle of an ever moving picture, driven himself by the irresistible torrent that carries all around him along, the American has no time to attach himself to anything, he is only accustomed to change and ends by looking on it as the natural state of man.'[23] The result was a paradox which is one of the central features of modern capitalism, that desire always outstrips achievement. 'At first sight there is something astonishing in this spectacle of so many lucky men restless in the midst of abundance.'[24]

Tocqueville was aware that in *ancien régime* societies, mercantile wealth tended to be looked down on. 'In aristocracies the rich are also the ruling class. Constant attention to great affairs of state diverts them from the petty cares of trade and industry. Should one of them nonetheless feel a natural inclination toward business, corporate public opinion at once bars his path.'[25] Likewise, manufacturing was vulgar and 'low-caste'. Beaumont, Tocqueville's travelling companion described a manufacturer thus: 'No elegance; good nature; polite; sometimes indiscreet; embarrassingly obliging; it's *absolutely America*.'[26] In other words, being 'in trade', whether as a merchant or manufacture was vulgar, vaguely dishonourable, somehow dirty and demeaning. But not in America.

The central American passion was the pursuit of profit, both as a means and as an end. Tocqueville comments on this with amazement. 'So one usually finds that love of money is either the chief or a secondary motive at the bottom of everything the Americans do.'[27] He noticed that 'A breathless cupidity perpetually distracts the mind of man from the pleasures of the imagination and the labours of the intellect and urges it on to nothing but the pursuit of wealth.'[28] He realized that this was partly to do with the 'open frontier' of America. 'To clear, cultivate and transform the huge uninhabited continent which is their domain, the Americans need the everyday support of an energetic passion; that passion can only be the love of wealth.'[29] But later, when he visited the French Canadians, he found that they lacked this mentality and realized that it was mainly cultural and historical, rather than caused by the vast 'emptiness' or the practicalities of battling with nature.

It continued to fascinate and surprise him. 'It is odd to watch with what feverish ardour the Americans pursue prosperity and how they are ever tormented by the shadowy suspicion that they may not have chosen the shortest route to get it.'[30] The desire for the shortest route was one of the reasons, he thought, for the huge inventiveness and conspicuously growing wealth of America.

They think about nothing but ways of changing their lot and bettering it. For people in this frame of mind every new way of getting wealth more quickly, every machine which lessens work, every means of diminishing the costs of production, every invention which makes pleasures easier or greater, seems the most magnificent accomplishment of the human mind.[31]

This tendency was made all the stronger by the huge size, diversity yet homogeneity, of America.

'In a large state thought on all subjects is stimulated and accelerated; ideas circulate more freely; the capitals are vast intellectual centres concentrating all the rays of thought in one bright glow; that is why great nations contribute more and faster to the increase of knowledge and the general progress of civilization than small ones.[32]

This mention of inventions and labour-saving devices is one of the few places where Tocqueville explicitly talks about the early stages of the technological and industrial revolution that had transformed England and were starting to do the same in America. As Schleifer points out, Tocqueville did, abstractly, recognize that the industrial revolution, along with the tendency to equality (democracy) was one of the great forces of his time.[33] He also showed some sporadic interest in specific technologies and industrialization.[34] But in the case of America, in one of his very few failures to see into the future, he predicted a great commercial, but not industrial, future for the country.[35]

The likelihood of continuing wealth accumulation was also heightened by what, to an aristocrat like Tocqueville, was a very strange attitude to work. He explained to his French contemporaries how it was in America. 'Among democratic peoples where there is no hereditary wealth, every man works for his living, or has worked, or comes from parents who have worked. Everything therefore prompts the assumption that to work is the necessary, natural, and honest condition of all men.'[36] Thus, 'Not only is no dishonour associated with work, but among such peoples it is regarded as positively honourable; the prejudice is for, not against, it.'[37] Honour, which lies in idleness in most societies, has been overturned. 'In a democratic society such as that of the United States, where fortunes are small and insecure, everybody works, and work opens all doors.

That circumstance had made the point of honour do an about turn and set it facing against idleness.'[38] Thus all occupations, as long as they make money, are honourable and the American is very versatile and flexible in his or her attitude. 'In the United States professions are more or less unpleasant, more or less lucrative, but they are never high or low. Every honest profession is honourable.'[39] People will often do several types of job, successively or simultaneously. 'In America it sometimes happens that one and the same man will till his fields, build his house, make his tools, cobble his shoes, and with his own hands weave the coarse cloth that covers him. This is bad for improving craftsmanship but greatly serves to develop the worker's intelligence.'[40]

Tocqueville at times implied that perhaps necessity was the mother of work, as it was of invention, that people worked so hard because of their small fortunes. Yet he realized that it was deeper than this. Even as they became wealthier, they were driven on. 'For them desire for well-being has become a restless, burning passion which increases with satisfaction.'[41] They exhibited a restrained, puritan passion for wealth.

Such passionate materialism started with the commercial middle classes, and spread out as that bourgeois group took over the heart of America and set its standards. 'The passion for physical comfort is essentially a middle-class affair; it grows and spreads with that class and becomes preponderant with it.'[42] It also spread out from the sphere of the economy into all of life. 'The passions that stir the Americans most deeply are commercial and not political ones, or rather they carry a trader's habits over into the business of politics.'[43] As Smith had earlier observed of England, it was a country 'ruled by shopkeepers'.

Those who have commented on Tocqueville have noted that he saw that *America* had somehow solved Adam Smith's contradiction between private desire and public benefit by harmonizing self-interest with public interest, creating a kind of calculative virtue.[44] As Lerner puts it, 'Time after time he confronts the paradox of a society which is fragmentised by self-interest and self-seeking but which seems nevertheless to have found a principle of inner order.'[45] Sometimes Tocqueville just recognizes that somehow this has been achieved. 'What a happy land the New World is, where man's vices are almost as useful to society as his virtues!'[46] At other times he points to the way astute politicians and lawyers frame their activities to bring public and private good together. Thus 'American legislation appeals mainly to private interest; that is the great principle which one finds again and again when one studies the laws of the United States.'[47] He noted furthermore that 'American legislators show little confidence in human honesty, but they always assume that men are intelligent. So they generally rely on personal interest to see to the execution of the laws.'[48]

In one of the few places where, in his travel journals, he tried to tackle the problem he wrote:

> The two great social principles which seem to me to rule American society and to which one must always return to find the reason for all the laws and habits which govern it, are as follows: 1st. The majority may be mistaken on some points, but finally it is always right and there is no moral power above it. 2nd. Every individual, private person, society, community or nation, is the only lawful judge of its own interest, and provided it does not harm the interests of others, nobody has the right to interfere. I think that one must never lose sight of this point.[49]

*

As for the consequences of the restless pursuit of profitable activity, Tocqueville notes several unexpected results. We have seen that it was combined with surprising restraint, not just as a result of the Puritan heritage. Paraphrasing certain themes in Montesquieu and Smith on the pacifying effects of the pursuit of wealth, he noted: 'Trade is the natural enemy of all violent passions. Trade loves moderation, delights in compromise, and is most careful to avoid anger.'[50] With an obvious message for his own revolution-prone country, and making a helpful distinction between permanent surface change, and the absence of fundamental revolutions, he wrote: 'Daily they change, alter and renew things of secondary importance, but they are very careful not to touch fundamentals. They love change, but they are afraid of revolutions.'[51]

The constant immersion in the pursuit of material goals also altered the whole attitude to time and the momentum of history. Time past was irrelevant. 'Aristocracy naturally leads the mind back to the past and fixes it in the contemplation thereof. But democracy engenders a sort of instinctive distaste for what is old.'[52] Tocqueville saw that political, social and physical time are interrelated, a sort of Einsteinian view of the relativity of concepts of time and social relations.

Among democratic peoples new families continually rise from nothing while others fall, and nobody's position is quite stable. The woof of time is ever being broken and the track of past generations lost. Those who have gone before are easily forgotten, and no one gives a thought to those who will follow. All a man's interests are limited to those near himself.[53]

He noted the optimism and future-orientation of the Americans.

Howsoever powerful and impetuous the course of history is here, imagination always goes in advance of it, and the picture is never large enough. There is not a country in the world where man more confidently takes charge of the future, or where he feels with more pride that he can fashion the universe to please himself.[54]

These shocks and surprises at the turbulence and commercial spirit of America led Tocqueville to ponder on how the system could work like this. This presented him with further puzzles. He could see that the ever-striving, hard-working and calculating spirit was somehow linked to the political system. He made a strong connection between political freedom and the generation of 'wealth' or wellbeing. He first noticed that this was a characteristic of 'democratic' countries. 'There is therefore at the bottom of democratic institutions some hidden tendency which often makes men promote the general prosperity, in spite of their vices and their mistakes, whereas in aristocratic institutions there is sometimes a secret bias which, in spite of talents and virtues, leads men to contribute to the afflictions of their fellows.'[55] On the basis of his later experience in England he widened this into a universal proposition. 'I doubt if one can cite a single example of any people engaged in both manufacture and trade, from the men of Tyre to the Florentines and the English, who were not a free people. There must therefore be a close link and necessary relationship between these two things, that is, freedom and industry.'[56] But the actual causal links were very difficult to discern.

At times he seemed to suggest that the bourgeois mentality was the most important, affecting political institutions and thence wealth. 'Everyone living in democratic times contracts, more or less, the mental habits of the industrial and trading classes; their thoughts take a serious turn, calculating and realistic; they gladly turn away

176 The Riddle of the Modern World

from the ideal to pursue some visible and approachable aim which seems the natural and necessary object of their desires.'[57] At other times he emphasized freedom and education and almost exactly paraphrased Adam Smith's 'peace, easy taxes and a tolerable administration of justice'.

> If you give democratic peoples education and freedom and leave them alone, they will easily extract from this world all the good things it has to offer. They will improve all useful techniques and make life daily more comfortable, smooth, and bland. Since their social condition by its nature urges them this way, there is no need to fear that they will stop.[58]

Another link was between the degree of political absolutism and centralization on the one hand and wealth creation on the other. He believed that

> It is certain that despotism brings men to ruin more by preventing them from producing than by taking away the fruits of their labours; it dries up the fount of wealth while often respecting acquired riches. But liberty engenders a thousandfold more goods than it destroys, and in nations where it is understood, the people's resources always increase faster than the taxes.[59]

One way in which this insidiously happened was through the draining of the more innovative from the countryside as centralization proceeded.

> Is it a centralized country? The rural districts are emptied of rich and enlightened inhabitants. I could go further – a centralized country is a country of imperfect and unprogressive cultivation; and I could comment on the profound saying of Montesquieu by explaining his meaning – 'lands produce less by reason of their fertility than by reason of the liberty of their inhabitants.'[60]

Perhaps the nearest he came to resolving the difficulty of reciprocal causation was when he wrote

> I have no doubt that democratic institutions, combined with the physical nature of the land, are the indirect reason, and not, as is often claimed, the direct one, for the prodigious industrial

expansion seen in the United States. It is not the laws' creation, but the people have learned to achieve it by making the laws.⁶¹

Here he recognized that there was something behind the laws – returning again to the primacy of culture.

Even if he had been content to explain the situation fully in terms of the legal and political 'freedom', he would have faced a serious problem. This was because, from a French standpoint 'America' seemed to run itself without any obvious political system at all. Again it seemed to have achieved the impossible, to be very well organized and orderly, with few signs of government.

When Tocqueville first arrived he expressed his astonishment at the bizarre situation.

> What is most striking to everyone who travels in this country, whether or not one bothers to reflect, is the spectacle of a society marching along all alone, without guide or support, by the sole fact of the cooperation of individual wills. In spite of anxiously searching for the government, one can find it nowhere, and the truth is that it does not, so to speak, exist at all.⁶²

How then was it held together?

One part of the solution lay in decentralized power, which, like Montesquieu, he admired. Yet he was fully aware that too much decentralization could be disastrous. Thus, looking back at the early mediaeval period in Europe he described it thus: 'the cause of all the miseries of feudal society was that power, not just of administration, but of government, was divided among a thousand people and broken up in a thousand ways; the absence of all governmental centralisation then prevented the nations of Europe from advancing energetically toward any goal.'

Tocqueville explained that the Americans had created an 'imagined community' to hold together, through ideology, an equal peoples who thus needed few police, no central bureaucracy, no standing army. He noted that, in contrast to France, 'one is bound to notice that all classes show great confidence in their country's legislation, feeling a sort of paternal love for it'.⁶³ Using 'ideal' in the sense of imagined, he wrote that 'The government of the Union rests almost entirely on legal fictions. The Union is an ideal nation which exists, so to say, only in men's minds and whose extent and limits can only be discerned by the understanding.'⁶⁴ This ideal community was highly artificial, manufactured, yet it felt 'natural'. He believed that it was only possible because it was founded on that most powerful set of American institutions, the self-governing commune, and a plethora of different institutions. 'Everything in such a government depends on artificially contrived conventions, and it is only suited to a people long accustomed to manage its affairs, and one in which even the lowest ranks of society have an appreciation of political science.'[65] The vitality was at the local level, and fed upwards. Centralized aristocracies like France were top-down, and the nation was only held together by physical force. 'So, whereas with us the central government lends its agents to the commune, in America the township lends its agents to the government. That fact alone shows how far the two societies differ.'[66]

The whole system depended on the dynamic creation and maintenance of 'artificial communities' at the lower levels. Tocqueville had earlier noted that 'The American people taken in mass is not only the most enlightened in the world, but – what I put much higher than that advantage – is the one whose practical political education is the most advanced.'[67] This practical education was absolutely essential for democracy to work and hence for wealth to increase, especially as a counterbalance to the dangers of narrow individualism generated by growing equality. 'If men are to remain civilised or to become civilised, the art of association must develop and improve among them at the same speed as equality of conditions spreads.'[68] And this is exactly what he found in America. 'Better use has been made of association and this powerful instrument of action has been applied to more varied aims in America than anywhere else in the world.'[69] He found that

> Americans of all ages, all stations in life, and all types of disposition are for ever forming associations. There are not only commercial and industrial associations in which all take part, but others of a thousand different types – religious, moral, serious, futile, very general and very limited, immensely large and very minute. Americans combine to give fetes, found seminaries, build churches, distribute books, and send missionaries to the antipodes. Hospitals, prisons, and schools take shape in that way.[70]

He had heard that the English were famous for their associations, or, as Montesquieu would have described them perhaps, 'intermediary institutions'. Yet while 'A single Englishman will often carry through some great undertaking', Tocqueville found that 'Americans form associations for no matter how small a matter. Clearly the former regard association as a powerful means of action, but the latter seem to think of it as the only one.'[71] This was one explanation for their dynamism for it made them self-confident, experienced in practical politics, unafraid of the State.[72] Indeed Tocqueville placed great emphasis on individual self-responsibility in government and other spheres and this was one of the facts which attracted him to the American legal system as a central feature in his explanation of democracy. Firstly, whereas in France the political institutions dominated the legal ones, in America it was the other way round. 'In a sense the legislature penetrates to the very heart of the administration.'[73] Secondly, the legislation itself was made for the people and not for the State and hence had a distinct flavour. 'Nothing is more peculiar or more instructive than the legislation of this time; there, if anywhere, is the key to the social enigma presented to the world by the United States now.'[74] Thirdly, the dominance of law was combined with delegation downwards. 'In no country in the world are the pronouncements of the law more categorical than in America, and in no other country is the right to enforce it divided among so many hands.'[75]

Among these 'hands' two particularly struck him. One was the office of justice of the peace. As he realized, this was a system that had been introduced from England. 'The Americans have borrowed from their English forefathers the conception of an institution which has no analogy with anything we know on the Continent, that of justices of the peace.'[76] The other major delegation of power had also been borrowed from England, namely the jury system. In a number of places Tocqueville explained how juries not only protected the individual citizen against the power of the State, but, perhaps even more importantly, involved them in responsibility for their own governance. It was the most important 'political' education they had and trained them to participate properly in a democracy.

*

There was one further area where Tocqueville believed he had found a key to American civilization. That was religion, in particular the *relations* of religion to the political system.[77] Tocqueville's account captures yet again the deep contradictions in the system. On the one hand, it was clear to him that religion was enormously important as a social glue and as a source of consolation. Faith and hope he thought were two of '"the most permanent and invincible instincts of human nature" because "each has a need to nourish some illusion"'.[78] Without religious belief, man was easily seduced into terrible excesses. Thus in relation to the French revolution, he wrote that the 'universal discredit into which all religious beliefs fell at the end of the eighteenth century exercised without doubt the greatest influence on the whole course of our Revolution; it distinguished its character.'[79] Thus he was relieved to find that in many ways America was far more full of genuine religious activity than the Europe he had left. 'It is evident that there still remains here a larger foundation for Christian religion than in any other country in the world, to my knowledge...'[80] Although he modified this a little in the published book, he still argued that 'In America religion is perhaps less powerful than it has been at certain times and among certain peoples, but its influence is more lasting. It restricts itself to its own resources, of which no one can deprive it.'[81] Yet he also recognized that it was a peculiar and different 'religion' to the Catholicism which he had rejected in France.

He noted the absence of a dominant religious authority and hence the freedom to follow reason and individual conscience. In parallel to the decentralization in politics, this led to decentralized religion where the 'sects' became the equivalents to companies (economics) or communes (politics). He described the situation and his bafflement early on in his stay. 'Thus you see: Protestantism, a mixture of authority and reason, is battered at the same time by the two absolute principles of *reason* and *authority*. Anyone who wants to look for it can see this spectacle to some extent everywhere; but here it is quite striking.'[82] There are obvious echoes of his own earlier battle between reason and authority. And this perhaps led to an early somewhat cynical observation which he later dropped – for although it captured the optional nature of particular faith it did not capture the sincerity. 'People follow a religion the way our fathers took a medicine in the month of May – if it does not do any good, people seem to say, at least it cannot do any harm, and, besides, it is proper to conform to the general rule.'[83]

Part of the mystery was resolved by seeing that religion had been separated from politics. 'European Christianity has allowed itself to be intimately united with the powers of this world.'[84] This had not happened in America: State and Church were separate. This was recognized as a cause of the mutual harmony of each. Thus in

America 'all thought that the main reason for the quiet sway of religion over their country was the complete separation of church and state'.[85] It meant that while political and economic life could be turbulent, religion could be calm and certain. 'Thus, in the moral world everything is classified, coordinated, foreseen, and decided in advance. In the world of politics everything is in turmoil, contested and uncertain.'[86] It was extraordinary, but it worked. By separating the two worlds of politics and religion, they came to support each other better than by forcing them into the kind of concordats he was familiar with in Europe. 'Far from harming each other, these two apparently opposed tendencies work in harmony and seem to lend mutual support.'[87] Not that this flowed from any intrinsic lack of zeal, or even lack of ambition on the part of the sects. It was more, as Adam Smith and others had argued, the result of stalemate. America showed this wonderfully 'because the religious and irreligious instincts which can exist in man develop here in perfect liberty.'[88]

Particularly interesting was the position of the Catholic clergy.

> Protestants of all persuasions – Anglicans, Lutherans, Calvinists, Presbyterians, Anabaptists, Quakers, and a hundred other Christian sects – this is the core of the population. This church-going and indifferent population, which lives day to day, becomes used to a *milieu* which is hardly satisfying, but which is tranquil, and in which the *proprieties* are satisfied. They live and die in compromises, without ever concerning themselves with reaching the depths of things; they no longer recruit anyone. Above them is to be found a fistful of Catholics, who are making use of the tolerance of their ancient adversaries, but who are staying basically as intolerant as they have always been, as intolerant in a word as people who *believe*.[89]

He tried to capture the same point with a metaphor of a set of concentric circles, with the Catholics in the middle.

> It is an incredible thing to see the infinite subdivisions into which the sects have been divided in America. One might say they are circles successively drawn around the same point; each new one is a little more distant than the last. The Catholic faith is the immobile point from which each new sect distances itself a little more, while drawing nearer to pure deism.[90]

In such a situation the extreme dogmatists, the Catholics, not only had to abandon any idea of an alliance with the State, but they were pushed back into a purely private role. Because of the multiplicity of sects and of different priests, each religious group became relegated to the level of the private life of the citizen. They had no choice but to accept that religion and their particular morality stopped at the front door of their sect follower; it could not be imposed on others. 'American Catholic priests have divided the world of the mind into two parts; in one are revealed dogmas to which they submit without discussion; political truth finds its place in the other half, which they think God has left to man's free investigation.'[91]

Thus religion was inwardly strong and outwardly weak.

> Religion in America is a world apart in which the clergyman is supreme, but one which he is careful never to leave; within its limits he guides men's minds, while outside them he leaves men to themselves, to the freedom and instability natural to themselves and the times they live in. I have seen no country in which Christianity is less clothed in forms, symbols, and observances than it is in the United States, or where the mind is fed with clearer, simpler, or more comprehensive conceptions.[92]

It was not a 'civil religion', but a privatized, individualized, yet heavily ethical world continuing the traditions of its Pilgrim fathers.

Like Montesquieu before him, Tocqueville seems to have realized that there was something about this religious structure which was particularly propitious for the development of what we would now call capitalism. It was not so much the actual dogma, but the structural position of religion and the spur it gave. Part of this is caught by Lerner when he writes: 'He even saw what Max Weber and R.H. Tawney were to see later: that there was an inner relation between the religious spirit and the strength of the capitalist impulse in America, and that the single-minded pursuit of wealth and personal property was linked with the single-minded quest of God.'[93] Or as Tocqueville himself put it, in a paradox of the same kind as Weber's 'That is why religious nations have often accomplished such lasting achievements. For in thinking of the other world, they had found out the great success in this.'[94]

*

Tocqueville synthesized all these ideas into one major theory: that the development of the spirit of equality was the key to American civilization. In America he found a land which had explicitly enthroned the premise of equality, rather than of inequality. It made it a central tenet that man was born free and equal. This was still a peculiar way to look at things and Tocqueville consequently noted that 'No novelty in the United States struck me more vividly during my stay there than the equality of conditions.'[95] Equality, or democracy as he often called it, became the key to understanding America. 'So the more I studied American society, the more clearly I saw equality of conditions as the creative element from which each particular fact derived, and all my observations constantly returned to this nodal point.'[96] There had been some early attempts to take inequality over from the Old World, but they had failed. 'Laws were made there to establish the hierarchy of ranks, but it was soon seen that the soil of America absolutely rejected a territorial aristocracy.'[97]

He became convinced after his visit that this growing equality was the future. 'It seems to me beyond doubt that sooner or later we, like the Americans, will attain almost complete equality of conditions.'[98] He believed that 'the gradual process of equality is somehow fated.'[99] Governments could channel its course, but not stop it. 'In a word, from now on democracy seems to me a fact that a government can have the pretension of *regulating*, but of stopping, no.'[100] This was all the more so because there was a positive feedback. The more equality there was, the more impatient people became at the remaining inequalities. 'When inequality is the general rule in society, the greatest inequalities attract no attention. When everything is more or less level, the slightest variation is noticed. Hence the more equal men are, the more insatiable will be their longing for equality.'[101]

His certainty did not only arise from his American experience. His increasing research into the history of Europe appeared to show the same tendency. When people suggested that a new aristocracy created by industrial or commercial wealth might reinstate hierarchy, Tocqueville was prepared to concede temporary, small-scale, reversals. 'Hence, just while the mass of the nation is turning toward democracy, that particular class which is engaged in industry becomes more aristocratic.'[102] Yet this was only marginal. 'Does anyone imagine that democracy, which has destroyed the feudal system and vanquished kings, will fall back before the middle classes and the rich?'[103] Thus his work on *Democracy in America* was impelled by a need to understand and direct this tendency. 'This whole book has been written under the impulse of a kind of religious dread inspired by contemplation of this irresistible revolution advancing century by century over every obstacle and even now going forward amid the ruins it had itself created.'[104]

Tocqueville's greatness lies in the fact that as a member of a noble family, he nevertheless partially rejected the premise of natural inequality which had been the foundation of his ancestral power. He realized that 'rational equality is the only state natural to man'.[105] This was proved not only by the fact that 'nations get there from such various starting points and following such different roads'[106] but also by the evidence that 'Running through the pages of our history, there is hardly an important event in the last seven hundred years which has not turned out to be advantageous for equality'.[107] His central dynamic, therefore, was the tendency towards equality, and the movement away from birth to achievement as the basis for social position. This was the unstoppable force, the tide of history: 'the great human revolution which we set in motion more than sixty-five years ago, advances towards liberty only occasionally, but towards equality with an irresistible and uninterrupted progress.'[108]

His anxiety was that such a tendency could lead either to the elevation or subjugation of men. 'To me the Christian nations of our day present an alarming spectacle; the movement which carries them along is already too strong to be halted, but it is not yet so swift that we must despair of directing it; our fate is in our hands, but soon it may pass beyond control.'[109] His central conclusion in the first volume of *Democracy* was summarized thus by J.S. Mill.

> They may be stated as follows: – That Democracy, in the modern world is inevitable; and that it is, on the whole, desirable; but desirable only under certain conditions, and those conditions capable, by human care and foresight, of being realized, but capable also of being missed. The progress and ultimate ascendancy of the democratic principle has, in his eyes, the character of a law of nature.[110]

*

Tocqueville was particularly interested in the effects of the advancing tide of equality on interpersonal relations. He noted the effects on parent–child relations. He wrote that 'Everyone has noticed that in our time a new relationship has evolved between the different members of a family, that the distance formerly separating father and son has diminished, and that paternal authority, if not abolished, has at least changed form.'[111] Thus, in America, 'the family, if one takes the word in its Roman and aristocratic sense, no longer exists.'[112] The phenomenon could be seen as a child grew up in America, for 'as soon as the young American begins to approach man's estate, the reins of filial obedience are daily slackened. Master of his thoughts, he soon becomes responsible for his own behaviour. In America there is in truth no adolescence. At the close of boyhood he is a man and begins to trace out his own path.'[113]

The independence of children and their separation from their parents could be seen when he compared his own aristocratic childhood with what he saw in America. In the former, patriarchal power was still present. He believed that 'When men are more concerned with memories of what has been than with what is, and when they are much more anxious to know what their ancestors thought than to think for themselves, the father is the natural and necessary link between the past and the present, the link where these two chains meet and join.'[114] On the other hand, in America 'When the state of society turns to democracy and men adopt the general principle that it is good and right to judge everything for oneself, taking former beliefs as providing information but not rules, paternal opinions come to have less power over the sons, just as his legal power is less too.'[115] Here, as elsewhere, hierarchy and holism were linked on one side, with equality and individualism as a matched pair on the other.

The change from patriarchal to egalitarian family structures was obviously connected to the change from a situation where the state used the father, to one which separated politics and the family. 'As in aristocratic society, so in the aristocratic family, all positions are defined. Not only the father holds a rank apart and enjoys immense privileges; the children too are by no means equal among one another.'[116] Thus

> In aristocracies society is, in truth, only concerned with the father. It only controls the sons through the father; it rules him, and

he rules them. Hence the father has not only his natural right. He is given a political right to command. He is the author and support of the family; he is also its magistrate.[117]

But this was not the case in democracies, where each member of the family was a free citizen, responsible for himself. The alteration in power was perceptible in the tone of the relationship of fathers and sons in America. For 'among democratic nations every word a son addresses to his father has a tang of freedom, familiarity, and tenderness all at once, which gives an immediate impression of the new relationship prevailing in the family.'[118] Thus the premise of equality changed all the relations between the generations, and even between older and younger brothers.

Just as the relations between the generations was deeply affected, so was the relation between the genders. Tocqueville believed this to be a very important topic. 'Therefore everything which has a bearing on the status of women, their habits, and their thoughts is, in my view, of great political importance.'[119] He wondered whether 'democracy' was likely to destroy or modify 'the great inequality between man and woman which has up till now seemed based on the eternal foundations of nature?'[120] Personally he felt sure that it would 'raise the status of women, and should make them more and more nearly equal to men'.[121] He then outlines the high status of American women. For instance,

> In Europe one has often noted that a certain contempt lurks in the flattery men lavish on women; although a European may often make himself woman's slave, one feels that he never sincerely thinks her his equal. In the United States, men seldom compliment women, but they daily show how much they esteem them.[122]

The Americans carried to an extreme a tendency which Tocqueville had noticed seemed somehow to be linked to Protestantism and liberty, and hence was also found in England.

> In almost all Protestant nations girls are much more in control of their own behaviour than among Catholic ones. This independence is even greater in those Protestant countries, such as England, which have kept or gained the right of self-government. In such cases both political habits and religious beliefs infuse a spirit of liberty into the family.[123]

The fact that Tocqueville was married to a middle-class English woman, Mary Mottley, gave him an especial insight into these cultural differences.

Tocqueville noted that the freedom of American women started when they were young. 'Long before the young American woman has reached marriageable age, the process of freeing her from her mother's care has started stage by stage. Before she has completely left childhood behind she already thinks for herself, speaks freely, and acts on her own.'[124] This was reinforced by the relatively late age at marriage. 'Precocious weddings hardly occur. So American women only marry when their minds are experienced and mature, whereas elsewhere women usually only begin to mature when they are married.'[125] It was then maintained by a division of labour between the sexes. Men and women were given different spheres, in recognition of their different abilities, but each was valued highly. The Americans 'consider that progress consists not in making dissimilar creatures do roughly the same things but in giving both a chance to do their job as well as possible. The Americans have applied to the sexes the great principle of political economy which now dominates industry. They have carefully separated the functions of man and of woman so that the great work of society may be better performed.'[126] 'To sum up, the Americans do not think that man and woman have the duty or the right to do the same things, but they show an equal regard for the part played by both and think of them as beings of equal worth, though their fates are different.'[127]

Thus again there was a paradox. In many respects American women were quite confined in their role. Yet they had the highest 'station' or status in the world. 'For my part, I have no hesitation in saying that although the American woman never leaves her domestic sphere and is in some respects very dependent within it, nowhere does she enjoy a higher station.'[128] It was another instance of a blend of religion, economy and society. 'The Americans are both a Puritan and a trading nation. Therefore both their religious beliefs and their industrial habits lead them to demand much abnegation on the woman's part and a continual sacrifice of pleasure for the sake of business, which is seldom expected in Europe.'[129] And it was upon the superior ability and status and intelligence of American women that the greatness of America was based. 'If anyone asks me what I think the chief cause of the extraordinary prosperity and growing power of this nation, I should answer that it is due to the superiority of their women.'[130]

188 *The Riddle of the Modern World*

*

Tocqueville was well aware of the implicit dangers of the New World which he was analysing. In general he was very fair and balanced in his appraisal. He saw much to admire and to praise as we have seen, and may also note his admiration for the high educational standards and the independence of mind. It was a mighty, largely tolerant, and ambitious nation. Yet, alongside the achievements he noted not only the possibility of despotism and loneliness, but other presently existing evils. One, which he castigated with bitterness, was slavery which appalled him on his southern travels.[131] Another was the destruction of the American Indians. A book could be written just about his poignant account of the tragic destruction then in its last phases.[132] Here we can only cite three examples from his extensive journals and writings. As he watched the last huddled bands of Indians he felt an inexpressible sadness. 'There was, in the whole of this spectacle, an air of ruin and destruction, something that savoured of a farewell that was final and with no return; no one could witness this without being sick at heart; the Indians were calm, but sombre and taciturn.'[133] He summarized the European impact thus. 'The Europeans, having scattered the Indian tribes far into the wilderness, condemned them to a wandering vagabond life full of inexpressible afflictions.'[134] As he saw only too clearly, they were faced with an impossible choice. 'From whatever angle one regards the destinies of the North American natives, one sees nothing but irremediable ills: if they remain savages, they are driven along before the march of progress; if they try to become civilised, contact with more civilised people delivers them over to oppression and misery.'[135] Already he could see an end to their way of life and even, the prophet that he was, an end to the wilderness.

His vision of the ecological destruction that took place over the following century is shown in a moving passage written during his 'fortnight in the wilderness'. 'In a few years these impenetrable forests will have fallen; the sons of civilization and industry will break the silence of the Saginaw; its echoes will cease; the banks will be imprisoned by quays; its current, which now flows on unnoticed and tranquil through a nameless waste, will be stemmed by the prows of vessels.' It was the imminent loss which added to the beauty. 'It is this idea of destruction, with the accompanying thought of near and inevitable change, that gives to the solitudes of America their peculiar character, and their touching loveliness.'[136]

Yet, at a deeper level, he thought that even the white settlers were themselves being destroyed, but by a more insidious disease. There is sadness in this observation too:

> 'the people is becoming enlightened, attainments spread, and a middling ability becomes common. The striking talents, the great characters, are rare. Society is less brilliant and more prosperous. These various effects of the progress of civilization and enlightenment, which are only hinted at in Europe, appear in the clear light of day in America. From what first cause do they derive? I do not yet see clearly.'[137]

Later, though puzzled, he saw the malaise more clearly. 'Why, as civilisation spreads, do understanding men become fewer? Why, when attainments are the lot of all, do great intellectual talents become rarer? Why, when there are no longer lower classes, are there no more upper classes? Why, when knowledge of how to rule reaches the masses, is there a lack of great abilities in the direction of society? America clearly poses these questions. But who can answer them?'[138]

Another insidious evil which he saw emerging was an effect of that very principle of the division of labour which Smith had elaborated. Alluding explicitly to Smith, Tocqueville asked: 'What is one to expect from a man who has spent twenty years of his life making heads for pins? And how can he employ that mighty human intelligence which has so often stirred the world, except in finding out the best way of making heads for pins?'[139] He expressed the thought thus. 'As the principle of the division of labour is ever more completely applied, the workman becomes weaker, more limited, and more dependent. The craft improves, the craftsman slips back.'[140] The brave new world of industrial civilization which he saw ahead was not one he unequivocally welcomed. The hugeness of America, with its physical size and in its growing population, presaged dangers. Those who had destroyed the wilderness and the Indians might be the heirs to a poisoned chalice. 'Great wealth and dire poverty, huge cities, depraved morals, individual egoism, and complication of interests are so many perils which almost always arise from the large size of the state.'[141]

Tocqueville had shown how the system seemed, amazingly, to work. But why was it like that? How had it come into being? Having diminished the importance of geography, and put 'laws' into perspective, he was not left with much else. For a while, he and Beaumont thought that the explanation for the large middle class and equality might lie in the inheritance system. But, as Pierson points out, this was a red herring and, though it appeared in the *Democracy*, could not get them very far.[142] There was only one area left, 'culture', that is *mores* and customs. So Tocqueville turned to an explanation of these, reverting to his methodical device of the 'point of departure'.

As Tocqueville thought more and travelled further he came to the conclusion that the secret of 'America' would not to be found in America, but in Europe, and particularly in England. He placed his final ideas on the subject in a footnote to his last work, the *Ancien Régime*. He put forward the general proposition that 'The physiognomy of governments can be best detected in their colonies, for there their features are magnified, and rendered more conspicuous. When I want to discover the spirit and vices of the government of Louis XIV, I must go to Canada. Its deformities are seen there as through a microscope.'[143] The same was true of the relationship of the United States and England. 'In the United States... the English anti-centralisation system was carried to an extreme. Parishes became independent municipalities, almost democratic republics. The republican element, which forms, so to say, the foundation of the English constitution and English habits, shows itself and develops without hindrance.'[144] This also helped to explain the homogeneity of the United States – 'there was a strong family likeness between all the English colonies as they came to birth.'[145]

He developed the idea of a germ, or seed, which shaped the colony but then took certain early tendencies further than in the homeland. 'I do not think the intervening ocean really separates America from Europe. The people of the United States are that portion of the English people whose fate it is to explore the forests of the New World...'[146] 'That portion' had taken the central feature, liberty, with them. 'At the time of the first immigrations, local government, that fertile germ of free institutions, had already taken deep root in English ways, and therewith the dogma of the sovereignty of the people had slipped into the very heart of the Tudor monarchy.'[147] They also took the separation of religion and politics. 'Most of English America was peopled by men who, having shaken off the pope's authority, acknowledged no other religious supremacy; they therefore brought to the New World a Christianity which I can only describe as democratic and republican; this fact singularly favoured the establishment of a temporal republic and democracy. From the start politics and religion agreed, and they have not since ceased to do so.'[148] These central features had been taken to their logical extreme.

> If it be true that each people has a special character independent of its political interest, just as each man has one independent of his social position, one might say that America gives the most perfect picture, for good and for ill, of the special character of the English race. The American is the Englishman left to himself.[149]

This was because of two significant facts. First, physical distance had created the practical necessity for self-governing local government institutions to take a greater burden than in the old country. From the very early days in the new colonies 'One continually finds them exercising rights of sovereignty; they appointed magistrates, made peace and war, promulgated police regulations, and enacted laws as if they were dependent on God alone.'[150] Secondly, the old hierarchical structure of England was not transferred. There was no extreme division of landed wealth, no aristocracy, no traditional gentry. In a sense America was an extension of that great middling part of English social structure, from the yeoman up to the successful manufacturer or merchant. The two extremes, the landless poor and the owners of huge estates, had vanished. As a result, 'All, from the beginning, seemed destined to let freedom grow, not the aristocratic freedom of their motherland but a middle-class and democratic freedom of which the world's history had not previously provided a complete example.'[151] This was the fascination of studying America. Out of old English elements it had shaped something new and unprecedented. Because of its short history it was possible to observe exactly how it had started and how evolved and to see the combination of seed and maturation. 'America is the only country in which we can watch the natural quiet growth of society and where it is possible to be exact about the influence of the point of departure on the future of a state.'[152]

At a specific level there were particular institutional transfers. 'Thus the flowering of local government in America flowed from the essential principle of the English polity. Transported at a single stroke far from the feudal remnants of Europe, "the rural parish of the Middle Ages became the New England township."'[153] Yet this was just one element of the generalized system of freedom, carried from England.

> The English who emigrated three centuries ago to found a democratic society in the wilds of the New World were already accustomed in their motherland, to take part in public affairs; they knew trial by jury; they had liberty of speech and freedom of the press, personal freedom, and the conception of rights and the practice of asserting them. They carried these free institutions and virile mores with them to America, and these characteristics sustained them against the encroachments of the state.[154]

In turn, again linking freedom and wealth, he found that 'The English colonies – and that was one of the main reasons for their prosperity – have always enjoyed more internal freedom and political independence than those of other nations.'[155]

Tocqueville thus became increasingly aware, even during his stay in America, that in order to understand the origins and causes of that extraordinary land he would have to continue his travels. The answer to his riddle lay in the 'point of departure', and that point seemed to lie in England. Thus shortly after he returned from America he made a five week trip to England. Then in May to September 1835 he made a longer visit to study the country in depth. England, in fact, became his second home and not only because he had married an English wife. It was finally in England rather than America that he found the answer to some of his riddles. America was the future, but in order to understand that future one must understand the past and present of a small island which had recently developed into the most powerful nation the world had ever known.

11
How the Modern World Emerged

Although Tocqueville's interest in American origins was a contributing factor in taking him to England, his interest in that country anticipated his voyage to America. In October 1828 he wrote a long summary essay on England based on the work of the historian Lingard, which Gargan rightly describes as 'brilliant'.[1] When he attended Guizot's lectures, he heard a good deal more about the constitutional and social differences of France and England which intrigued him and deeply influenced his later interpretation.

When Tocqueville finally arrived in England in August 1833, for a first brief visit of five weeks, he was initially confused and in some ways disappointed. Part of the disappointment was social. He confessed to 'a continual dizziness and a profound feeling of my nullity. We were a great deal in America, we are hardly anything in Paris, but I assure you that it is necessary to go to below zero and to use what mathematicians call negative numbers to compute what I am here.'[2] A second disappointment was that at first sight his hunch that America was England writ large seemed not to be the case.

Tocqueville and Beaumont had planned to 'return to France by way of England' from America. They were unable to do so but, as Beaumont put it in 1833, they had hoped to find out what heritage "John Bull, father of Jonathan" had transmitted to his son.'[3] Yet when Tocqueville arrived in England he found that 'I am no longer in America.'[4] Indeed, as Drescher puts it, 'Nothing struck him more than the difference between the two societies.' "Nowhere," he observed, "do I find our America." '[5] Above all he seemed to find that, while America was based on the premise of equality, England was still a deeply 'aristocratic' rather than 'democratic' society in terms of its class structure – indeed in some ways more so than France.

> The position that fortune joined to birth gives here appears to me to be still a million feet above all the rest. You are aware that I cannot yet speak of the spirit of the English people: what I can say, what strikes me most up to the present in its mores, is their aristocratic exterior. The aristocratic spirit appears to me to have descended into all classes; *every marquis wants to have pages*, make no mistake about it. In short, I do not recognize our America here in any point.[6]

Yet what at first came as a disappointment turned out to be a great advantage. Instead of a simple contrast between *Ancien Régime* Europe on the one hand, and Anglo-American civilization on the other, Tocqueville was forced to consider a third case, overlapping with both France and America, yet very different from either. This provided a three-way comparison which helps give his speculations far greater depth and subtlety. Like Montesquieu before him, he found in England a world very different from France, and one which gave him hope. '"It is the greatest spectacle in the world, though all of it is not great," he wrote. "One encounters, above all, things unknown in the rest of Europe – things which consoled me."'[7] As Drescher notes, it seemed to him to contain both the old world and the new in almost equal measure and to stand on the exact intersection. 'With a pattern peculiar to itself, England seemed to contain so many elements of both social conditions that none could say whether it had not already crossed the invisible boundary.'[8]

Tocqueville never wrote a great book on England, like his *Democracy in America*. This is one of the reasons why his thoughts on that country have been for so long time over-shadowed by his writing on America and his work apparently devoted to France in the *Ancien Régime*. In fact, as Drescher's excellent *Tocqueville in England* shows, England was as important a 'thought experiment' for Tocqueville as was America: 'The British Isles were the source of some of his greatest insights, especially into the historical connection between the rise of democracy and the extension of bureaucratic centralization.'[9] His experiences in England 'gave him a comparative basis for a theory of the relation of ideas to social change, of the causes of and antidotes for revolutions'.[10] In particular his second, longer, visit of 1835 stimulated him immensely. As Drescher writes, 'the spoils of the eleven-week expedition were immense. Tocqueville and Beaumont had undertaken a complete revaluation of what aristocracy and the democratic revolution meant in their English context.'[11] Thus 'When Tocqueville and Beaumont left England early in July 1835, it marked the end of the fullest experience of their lives.'[12] It is clear that the stimulus was not just intellectual but also moral and emotional.[13] We have seen that Tocqueville often felt isolated in his thought. In England and especially with his English friends and English wife he found support for his new evaluation of the world. As Tocqueville himself wrote to an English friend, 'So many of my sentiments and ideas are English, that England has become intellectually my second country.'[14]

It was a country which, in another way, could provide a model for France. It had undergone the immense urban and industrial revolutions, yet not had to suffer the torment of continuing political revolutions. As he revisited it over time, he was constantly surprised how it managed to change and yet to remain the same. Thus when he revisited it in 1857, after revolutions had swept across Europe, 'English society surprised him by being so consistent with its old pattern. It appeared that if England had changed it was in reverse – that she was now even less agitated by revolutionary passions than in 1835.'[15] America had many advantages in its newness. England had achieved the even more difficult route from a mediaeval to a modern society without needing a political revolution of the kind that had occurred elsewhere.[16]

The English case, with all its obvious success as the greatest technological and military power in the world, provided Tocqueville with a rod with which to beat his fellow countrymen. His last great book on the *Ancien Régime* could not have been written without the English counter-case, which became far more important than the American one. His indictment of France was severe. 'No Frenchman could put down the "Ancien Régime" without noticing with equal terror that its author had assailed almost every class, every institution, and every event in recent French history.'[17] No wonder he felt morally isolated, for he also loved France. He was able to make his powerful attack only because of his emotional and intellectual contacts with England.

Once again following Montesquieu, Tocqueville increasingly saw beneath the surface and realized that England was an object lesson. His illuminating account now seems self-evident. Yet it was less obvious in the 1820s as he developed his thoughts. 'England itself, poorly known in any event, had not yet furnished the striking arguments in favour of liberty that it has since done. Free institutions produced internal and unseen effects which were hidden to foreigners; their fecundity and their greatness were not yet manifest.'[18] His interest in the country was increased by two further facts. Just as America was a case where one could watch England spreading out in a new space, likewise England was a place where one could see the European revolutions rippling out in a new environment. As he put it in a letter, referring to the English Civil War period for instance,

> The previous revolutions that the English have undergone were essentially English in *substance* and in *form*. The ideas that gave birth to them circulated only in England; the form in which these ideas clothed themselves was unknown to the continent; the means that were used in order to make them victorious were the product of mores, habits, laws, practices different or contrary to the mores, the habits, and laws of the rest of Europe (all of that up to a certain point). Those previous revolutions in England thus were an object of great curiosity to the philosophers, but it was difficult for them to give rise to a popular book among us. It is no longer so today: today it is the European revolution that is being continued among the English, but it is being continued there by taking wholly English forms.[19]

The second reason for being interested in England was because of its growing Empire, and particularly its increasing dominance in India. As he wrote in 1840, 'Nothing is less well known in France than the causes that produced and that sustain the astonishing greatness of the English in India. This subject, which has always been interesting, is wonderfully so now that all the great affairs of Europe have their centre in Africa.'[20] He decided to work on the subject and wrote in 1841, 'My intention is to occupy myself with India...'[21] He worked hard for two years. As it was, he gave up this projected book in December 1843 under the pressure of other work. We are left with a few hints of his attraction to the subject and an unmeasurable influence on his *Ancien Régime* when he discusses 'caste'.

Thus England was important to Tocqueville at many different levels. It might contain the secret of the extraordinary New World. It held the key to successful imperialism. It had somehow moved from ancient to modern without the trauma of anything analogous to the French Revolution. It had industrialized and urbanized two generations before anywhere else. And it espoused those values of liberty which he cherished. For all these reasons he devoted much of his thought to the country, though his insights are scattered through his letters, journals, and unpublished papers and in asides in his major works. What did he find?

Despite the Revolution, France was still largely an *ancien régime* country in Tocqueville's childhood. That it to say, it was still divided into the four estates of *paysans, bourgeois, clergé* and *nobilité*, even if this was officially below the surface. It was still largely an agrarian country, with pockets of commerce. It is because of this background that Tocqueville, like Montesquieu almost exactly a century before him, felt a sense of shock and otherness when he went to England in 1833. Indeed he approvingly quoted Montesquieu's remark that 'I am here in a country which hardly resembles the rest of Europe'.[22] Though France had changed enormously in the century since Montesquieu, England had changed equally fast, not through political revolution but through the socio-economic transformation of the Industrial Revolution and the widening of the franchise in 1832. If France was one end of the continuum, England was in the middle, something that needed to be understood as a bridge between old Europe and the new world of America.

*

One thing that struck Tocqueville was the general affluence of England. 'A Frenchman on seeing England for the first time is struck by the apparent comfort and cannot imagine why people complain.'[23] He found 'a nation among whom the upper classes are more brilliant, more enlightened and wiser, the middle classes richer, the poor classes better off than anywhere else'.[24] In contrast, writing in 1857 about the 1820s and 1830s, Tocqueville gives a picture of France as it was then, at least in the countryside.

> Thirty years ago the peasant was dressed in linen all the year round; now, the poorest family wears warm and substantial woollens. Then he ate black bread; his bread now would have appeared a luxury even to the rich of those days. Butcher's meat was then almost unknown. Twenty-five years ago the little town

of St. Pierre had only a single butcher: he killed a cow once a week, and had great difficulty in selling his meat. Now there are nine, and they sell more in a day than was then sold in a week. Nor is this peculiar. I have observed a similar change in Touraine, in Picardy, in all the Ile-de-France, and in Lorraine.[25]

Yet, despite the improvement of conditions in France, England's industrial progress was such that by the mid-century Tocqueville could write 'is there any single country in Europe, in which the national wealth is greater . . . society more settled and more wealthy?'[26] What then were the reasons for this relative wealth?

Of course the possession of India helped. 'It is an inexhaustible resource for it, all the more because the climate is so deadly that the odds are three to one that an Englishman will die there; but if he does not die, he is *sure* of getting rich.'[27] That England's agricultural system was 'the richest and most perfect in the world' was likewise important.[28] It was also obviously becoming the workshop of the world. Tocqueville knew that 'We live in a century, not of monasteries, but of railways and Exchanges'.[29] He visited Manchester and Birmingham, and described the latter as 'an immense workshop, a huge forge, a vast shop.'[30] He realized that 'manufacture and trade are the best-known means, the quickest and the safest to become rich'.[31] The English understood this. 'Newton said that he found the world's system by thinking about it the whole time. By doing the same, the English have got hold of the trade of the whole world.'[32] Yet Tocqueville realized that all these explanations themselves needed an explanation.

The English obsessions with money-making was particularly important for he had already found the same phenomenon in America. In England also 'all the resources of the human spirit are bent on the acquisition of wealth'.[33] He observed that 'In all countries it is bad luck not to be rich. In England it is a terrible misfortune to be poor. Wealth is identified with happiness and everything that goes with happiness; poverty, or even a middling fortune, spells misfortune and all that goes with that.'[34] Everything was permeated with monetary values. 'Intelligence, even virtue, seem of little account without money. Everything worthwhile is somehow tied up with money. It fills all the gaps that one finds between men, but nothing will take its place.'[35] All of men's powers were attracted towards it. 'In a nation where wealth is the sole, or even the principal foundation of aristocracy, money, which in all society is the means of pleasure, confers power also. Endowed with these two advantages, it succeeds in attracting towards itself the whole imagination of man.'[36] The same was true in America. It was not just the growing towns where the commercial passion ruled. Although it was a vast, largely agricultural, land, 'the Americans carry over into agriculture the spirit of a trading venture, and their passion for industry is manifest there as elsewhere.'[37] Even in the remotest parts of the apparent wilderness, where you might expect illiterate peasants, 'In these so called villages you find none but lawyers, printers and shopkeepers.'[38]

Tocqueville at times suggested that there had been a change to this attitude quite recently, in the late eighteenth century in England. 'Fifty years ago, more or less, this was an accomplished revolution in England. Since that time birth is but an ornament of, or at most a help towards getting wealth. Money is the real power.'[39] At other times he argued that the drive towards economic acquisition was much older – and hence its enormous effects visible all over the world. 'Take into account the progressive force of such an urge working for several centuries on several millions of men, and you will not be surprised to find that these men have become the boldest sailors and the most skillful manufactures in the world.'[40]

All this obsession with accumulating wealth seemed peculiar to a French nobleman. How much more so to the vast majority of mankind who had lived outside the market economy. When Tocqueville came into contact with the native Americans, 'I determined to have recourse to their cupidity. But there is no such a philosopher as the Indian. He has few wants, and consequently few desires. Civilization has no hold over him. He neither knows, nor cares for its advantages.'[41] Such a person

> smiles bitterly when he sees us wear out our lives in heaping up useless riches. What we term industry he calls shameful subjection. He compares the workman to the ox toiling on in a furrow. What we call necessaries of life, he terms childish playthings, or womanish baubles. He envies us only our arms.[42]

So if the new man of England and America was obsessed with wealth acquisition, why was this? The direction to look towards, Tocqueville believed, was the social structure and the political system. This obsession with wealth was the result of numerous intersecting forces, and among the most important was the insecurity and restlessness generated by the absence of a fixed social hierarchy and by a competitive and balanced political system. The restlessness and dynamism was European, and Tocqueville, as we have seen, felt it himself. Thus the preparations for an invasion of China, he thought was an example of 'European restlessness pitted against Chinese unchangeableness,'[43] but the restlessness was most extreme where, as in the 'perpetual restlessness of the Americans',[44] there was least formal hierarchy, political or social. The fear of failure, the constant insecurity and ambition, he saw as follows. 'In democratic countries, not matter how rich a man is, he is almost always dissatisfied with his fortune, because he finds that he is less wealthy than his father was, and he is afraid that his son will be less wealthy than he.'[45]

In such 'democracies', that is to say in societies where wealth and power were based on achievement rather than on birth, people were involved in a vast competitive gambling match. Fortune's wheel was constantly turning, opportunities to rise and fall abounded. The vision of Adam Smith had come to pass and the zestful pursuit of wealth had developed into a passionate, restless and never-ending game which contributed to the wealth of the nation.

> Chance is an element always present to the mind of those who live in the unstable conditions of a democracy, and in the end they come to love enterprises in which chance plays a part. This draws them to trade not only for the sake of promised gain, but also because they love the emotions it provides.[46]

Thus the answer to the puzzle of English and American wealth and dynamism seemed to lie in the area of politico-social structures. Echoing Montesquieu, Tocqueville thought that England's wealth and security 'does not flow from the goodness of all the individual laws, but from the spirit which animates the complete body of English legislation. The want of perfection in certain organs is no impediment, because its spirit throbs with life.'[47] In fact, there was a circular causation. Commercial wealth was both a cause and consequence of the instability and dynamism.

The commercial nature of England meant that wealth could be acquired from sources other than land and hence a parallel 'aristocracy' was constantly emerging and challenging the older families. 'In this way an aristocracy of wealth was soon established and, as the world became more civilised and more opportunities of gaining wealth presented themselves, it increased, whereas the old aristocracy, for the same reasons, continually lost ground.'[48] The consequences were status competition and uncertainty, a constant preoccupation with small marks of difference and attempts to out-do others. Paradoxically this meant that in the middle of the nineteenth century, England was more snobbish than France. 'The French wish not to have superiors. The English wish to have inferiors. The Frenchman constantly raises his eyes above him with anxiety. The Englishman lowers his beneath him with satisfaction.'[49]

Ranks still existed in England, but they were confused. 'When birth alone, independent of wealth, decides a man's class, each knows exactly where he stands on the social ladder. He neither seeks to rise nor fears to fall.' But 'when an aristocracy of wealth takes the place of one of birth, this is no longer the case.'[50] This is because

> As a man's social worth is not ostensibly and permanently fixed by his birth, but varies infinitely with his wealth, ranks still exist, but it cannot be seen clearly at first sight by whom they are represented. The immediate result is an unspoken warfare between all the citizens. One side tries by a thousand dodges to infiltrate, in fact or in appearance, among those above them. The others are constantly trying to push back these usurpers of their rights. Or rather the same man plays both parts... Such is the state of England today...[51]

One aspect of the difference which particularly struck Tocqueville was the difference of mentality between the French peasant and the English and American small rural farmer. Tocqueville gave an account of the French and German peasantry in the eighteenth century. For instance, 'as in our own day, the peasant's love for property in land was extreme, and all the passions born in him by the possession of the soil was aflame'.[52] This was totally different from England.[53] In England, 'land is a luxury; it is honourable and agreeable to possess it, but it yields comparatively little profit. Only rich people buy it.'[54]

As for America, it was like England. Tocqueville concluded after his many travels that 'there are no peasants in America.'[59] What he meant is shown when he described his journey into the wildest parts of America. In the most remote region, you think that you have at last reached the abode of an American peasant: you are wrong. You enter his hut, which looks the abode of misery; the master is dressed as you are; his language is that of the towns. On his rude table are books and newspapers; he takes you hurriedly aside to be informed of what is going on in Europe, and asks you what has most struck you in his country. He will trace on paper for you the plan of a campaign in Belgium.[56]

*

Tocqueville was impressed by the freedom and balance of the politico-legal system in England and even more so by its offspring in America. Alluding to Montesquieu's thoughts on England, he wrote that 'Their *constitution* was famous already and was thought to be different from that of other countries.'[57] He felt that 'Nowhere else in Europe as yet was there a better organised system of free government. No other country had profited so much from feudal organisation.'[58] And when all this was taken to America it shed much of the snobbery and many of the contradictions of the mother country and settled for a purer form of commercial, middle-class orientation. It was both a continuity and a transformation. 'In the North, the English background was the same, but every nuance led the opposite way... the two or three main principles now forming the basic social theory of the United States were combined... Their influence now extends beyond its limits over the whole American world.'[59] Above all it refined and strengthened the freedom and self-rule brought from the old country. When he visited America, Tocqueville wrote that 'The two things that I chiefly admire here are these: First, the extraordinary respect entertained for law: standing alone, and unsupported by an armed force, it commands irresistibly.'[60] This he believed was because 'they make it themselves and are able to repeal it'. The 'second thing for which I envy these people is, the ease with which they do without being governed. Every man considers himself interested in maintaining the safety of the public and the exercise of the laws. Instead of depending on the police he depends on himself.'[61]

He found these trends in England also.

> When I see the force given to the human spirit in England by political life, when I see the Englishman sure of the aid of his

laws, relying on himself and seeing no obstacle but the limits of his own powers, acting without constraint... animated by the idea that he can do everything... seeking the best everywhere; when I see him thus, I am in no hurry to observe whether nature has carved out ports around him, and given him coal and iron. The cause of his commercial prosperity is not there; it is within himself.'[62]

How was one to explain this? Here Tocqueville rejected race. When he visited Switzerland with its famed republic, he concluded 'The kingdom of England is a hundred times more republican than this republic. Others would say that this results from the differences in the races. But that is an argument that I will never admit except at the last extremity, and when there remains absolutely nothing to say.'[63] So how could it be explained? Obviously the key must lie in the particular histories of different nations over the last five hundred years. One aspect of this could be seen in the contrast between the English and French revolutions.

Tocqueville summarized his views forcefully to an English correspondent, Lady Theresa Lewis, who had written a book on Lord Clarendon's contemporaries. 'Your biographies show the truth of your remark, that no two things can be more unlike than your Revolution of 1640 and ours of 1789. No two things, in fact, can be more unlike than the state of your society and of ours at those two periods.'[64] He continued that 'These differences, added to those between the character and the education of the two nations, are such that the two events do not admit of even comparison.'[65] In England, the dispute was between two segments of the ruling elite. 'They were divided; they were opposed to one another, and they fought; but never, for a single day, did they abdicate.'[66] Whereas in France there was a real ideological and class revolution. The results of this difference could be seen, for instance, in the nature of the events. Tocqueville's father's hair had turned white and his mother had become the neurotic character whose anxiety had overshadowed his youth as a result of the Terror. In England, in contrast, the fact that the ruling classes were always in control meant that 'The consequences were, less boldness of intention, less violence of action, and a regularity, a mildness, even a courtesy, admirably described by you, which showed itself even in the employment of physical force.'[67] In France there was a great rupture, a real turning over or revolution, a change in the rules. In England there had been Clarendon's 'Great Rebellion' where one power group replaced another temporarily but the rules were not changed at a deep, structural, level.

Tocqueville greatly admired the harmony and freedom of thought in England. The 'union of all the educated classes, from the humblest tradesman to the highest noble, to defend society, and to use freely their joint efforts to manage as well as possible its affairs'. 'I do not envy the wealth or the power of England, but I envy this union. For the first time, after many years, I breathed freely, undisturbed by the hatreds and the jealousies between different classes, which, after destroying our happiness, have destroyed our liberty.'[68]

Yet this union had the effect of leading to that turbulence and public confrontational behaviour which Montesquieu had noticed as a necessary feature of democracy. Thus Tocqueville noted:

> No people carry so far, especially when speaking in public, violence of language, outrageousness of theories, and extravagance in the inferences drawn from those theories. Thus your A.B. says, that the Irish have not shot half enough landlords. Yet no people act with more moderation. A quarter of what is said in England at a public meeting, or even round a dinner table, without anything being done or intended to be done, would in France announce violence, which would almost always be more furious than the language had been.[69]

This was a real gap between the English (and Americans) and everyone else. 'There is one point in which the English seem to me to differ from ourselves, and, indeed, from all other nations, so widely, that they form almost a distinct species of men. There is often scarcely any connection between what they say and what they do.'[70] The result was, once again, the confusion of foreigners who were misled by English irony or apparent hypocrisy.

Tocqueville's balance between praise and criticism also comes out in his assessment of English justice. On the positive side 'My general impression is that English procedure is much more expeditious than ours; that it often excludes incriminating evidence; that the system of "examination and cross examination" is better than ours for petty cases; that the position of the accused would be infinitely better than in France.'[71] On the other, 'It is impossible to imagine anything more detestable than the criminal investigation police in

England.'[72] On the one hand it is a country where the citizen is safe from absolutist power. In France,

> Aided by Roman law and by its interpreters, the kings of the fourteenth and fifteenth centuries succeeded in founding absolute monarchy on the ruins of the free institutions of the middle ages. The English alone refused to adopt it, and they alone have preserved their independence.[73]

On the other hand, speaking of England, 'There is not a country in the world where justice, that first need of peoples, is more the privilege of the rich.'[74] The backbone of the system 'is administered by the Justices of the Peace who are nominated by the King'.[75] This is a source of strength and independence, but there is also a danger that unelected and unaccountable individuals will gain too much power, for as Lord Minto warned Tocqueville, the administrative system 'rests almost entirely on the Justices of the Peace, magistrates who are responsible to no one and are not paid for the performance of their duties'.[76]

In fact he had noted another paradox. The English judicial system was confused, unprincipled, inefficient and cumbersome. Yet it somehow protected the citizen against the State better than anywhere else in the world. 'English law may be compared to the trunk of an old tree on which lawyers have continually grafted the strangest shoots, hoping that though the fruit will be different, the leaves at least will match those of the venerable tree that supports them.'[77] On the surface the French system of law appeared greatly superior. If one looked at English law,

> Here are astounding defects. Compare this old-fashioned and monstrous machine with our modern judiciary system, and the contrast between the simplicity, the coherence, and the logical organisation of the one will place in still bolder relief the complicated and incoherent plan of the other. Yet there does not exist a country in which, even in Blackstone's time, the great ends of justice were more fully attained than in England; nor one where every man, of whatever rank, and whether his suit was against a private individual or the sovereign, was more certain of being heard, and more assured of finding in the court ample guarantees for the defence of his fortune, his liberty, and his life.[78]

Tocqueville ended up by commending the English system with its Gothic extravagances and deep inconsistencies as far superior to his own sleeker system.

> Studying the judiciary system of England by the light of this principle, it will be discovered that while defects were allowed to exist which rendered the administration of justice among our neighbours obscure, complicated, slow, costly, and inconvenient, infinite pains had been taken to protect the weak against the strong, the subject against the monarch; and the closer the details of the system are examined, the better will it be seen that every citizen had been amply provided with arms for his defence, and that matters had been so arranged as to give to everyone the greatest possible number of guarantees against the partiality and venality of the courts, and, above all, against that form of venality which is both the commonest and the most dangerous in democratic times – subserviency to the supreme power.[79]

Thus the judicial system, as he had already argued in relation to America, was at the heart of the freedom of the English. The other central protection was the degree of decentralized power.

Beaumont and Tocqueville soon noted the importance of local institutions in England.

> 'One must go to the meetings of a Vestry,' wrote Beaumont, 'to judge what extraordinary liberty can be joined to inequality. One can see with what independence of language the most obscure English citizen expresses himself against the lord before whom he will bow presently. He is not his equal, of course, but within the limits of his rights he is as free, and he is fully aware of it.'[80]

The vestry was but one of the numerous local associations which were important and made each parish an almost self-governing community. As Drescher puts it, quoting Tocqueville and then adding his summary:

> 'The ensemble of English institutions is doubtless an aristocratic government, but there is not a parish in England which does not constitute a free public.' The parish, then, was the fundamental unit of public participation, the centre of a multitude of

interests vital to everyone in the community. For Tocqueville it was a complete democracy at the base of the social edifice.[81]

We are told that

> In his notes, Tocqueville wrote that if he were a friend to despotism, he would allow 'the deputies of the country [to deliberate] freely about peace and war, about the nation's finances, about its prosperity, its industries, its life. But I would avoid agreeing, at any price, that the representatives of a village had the right to assemble peacefully to discuss among themselves repairs for their church and the plan for their parsonage.'[82]

This led Tocqueville on to one of his greatest themes – the need for a balance between centralization and de-centralization. He was convinced that it was here, ultimately, that the secret of England and America's greatness must lie. Speaking of England he wrote:

> In that country the system of decentralisation, restricted from the beginning to proper limits, has attached to it nothing but notions of order, prosperity, and glory. The system of decentralisation has made, and still makes, the strength of England. England has had strong despotic kings at a time when the kingship was too primitive to want to undertake everything. The kings established centralisation of government; morals and the state of society caused administrative decentralisation.[83]

*

After his examinations of the political, social and legal factors which encouraged liberty, Tocqueville tended to come up with a version of the same theme that we have seen in Montesquieu and Smith, that a characteristic of the modern world, and one which distinguishes it from earlier civilizations, is that political and religious freedom seem to have a close association with the generation of economic wealth through the production of artefacts.

He believed that 'Geographical position and freedom had already made England the richest country in Europe'.[84] Freedom was necessary because, as Smith had argued, it allowed the acquisitive urges to fulfil themselves. 'Freedom in the world of politics is like the air in the physical world. The earth is full of a multitude of beings differently organized; but they all live and flourish. Alter the condition of the air, and they will be in trouble.'[85] Thus one should, in estimating the likelihood of wealth, 'Examine whether this people's laws give men the courage to seek prosperity, freedom to follow it up, the sense and habits to find it, and the assurance of reaping the benefit'.[86] That assurance of reaping the profit was equally important. Like Montesquieu and Smith, he realized that political and legal security, and in particular the safeguarding of a person's assets against the vagaries of war, arbitrary taxes and capricious law, were essential. The outstanding English encapsulation of this security was in their security of property. The extensive national and individual wealth of the English he linked to the fact that such wealth was 'more secure' than anywhere else.[87] This was linked to private property, 'exclusive proprietorial jealousy being so far developed here that it counts as one of the main national characteristics.'[88] That same English spirit had been carried to its overseas Empire. 'The English colonies – and that was one of the main reasons for their prosperity – have always enjoyed more internal freedom and more political independence than those of other nations.'[89]

*

Tocqueville had explained in what ways England was already a 'modern' country when it colonized America. He had shown that it was by the seventeenth century very different in its basic structure from most continental countries, especially France. Yet this left a further set of puzzles, in particular the dating of the divergence from Europe and the reasons for that parting of the ways. Essentially his answer to these questions is an expansion, with the historical sources carefully checked, of Montesquieu's argument. He suggested that out of a common European feudalism, that is the odd mixture which arose out of a decomposing Roman civilization and Germanic customs, the subsequent trajectory of continental Europe and England was different.

Tocqueville started with the premise that there had been very little difference between the parts of western Europe in the Dark Ages. The system which emerged in about the ninth century covered the whole of western and central Europe. 'If the feudal system is due to chance in France, by what odd coincidence does it turn up again among the Germans, among the Poles where it still exists, among the Goths in Spain, and even in Italy, the Southern extremity of Europe?'[90] It was already established in principle well before the eleventh century, and thus the best place to study it was in the earliest Saxon and Danish laws. Thus '... if you want to understand the first underlying principles of the feudal system, and you need to understand them to see how the wheels work in the finished machine, you cannot do better than study the time before the Norman conquest, because, as I said before, we know of no people nearer to their primitive state than the Saxons and the Danes.'[91] Many of these ancient principles never disappeared and, paraphrasing Montesquieu's famous remark about the origins in the German woods, Tocqueville thought that 'the customs of the Saxons are interesting in themselves and especially interesting in the context of English history. Their legal procedure is the oddest which has ever existed, and one can find in it all the elements of the present-day procedure, some parts of which we have adopted ourselves.'[92]

Then came the invasion of England by Tocqueville's Norman ancestors. William the Conqueror and his successors were able to lay out a complete 'system' so easily because they were merely codifying what was already there. 'Clearly the feudal system of the twelfth century is but the result of an underlying cause. It sprang fully armed from the peoples of the North, like Minerva from the head of Jupiter, needing only the hatchet's blow.'[93] At this point, Normandy and much of France, as well as most of the Continent, were identical to England. 'In comparing the feudal institutions in England immediately after the conquest with those in France, you find between them not only an analogy, but a perfect resemblance... In reality, the system in the two countries is identical.' However this identical system produced contrary results. He notes that Macaulay in his *History of England* 'alludes to the fact that England developed an open class structure, and France developed closed "castes", but he does not try to explain it'. Yet, why this divergence occurred is the key question, for there is no other which would provide 'so good an explanation of the difference between the history of England and that of the other feudal nations in Europe'.[94]

Tocqueville's conviction concerning the important difference between English social structure and that of the Continent, and his puzzlement as to why it should have occurred, continued in his *L'Ancien Régime* of 1856. He starts with the same assertion of a common starting point. 'I have had occasion to study the political institutions of the Middle Ages in France, in England, and in Germany, and the greater progress I made in this work, the more was I filled with astonishment at the prodigious similarity that I found between all these systems of law...' Thus 'in the fourteenth century the social, political, administrative, judicial, economic, and literary institutions of Europe had more resemblance to each other than they have perhaps even in our own days...'[95] He was struck by the fact that 'At that time many of the episodes of our history look as if they were drawn from the history of England. Such events never occurred in the following centuries.'[96] His picture is again one of divergence from a common origin. Starting with the thirteenth century,

> At this time there were to be found, as I have already said, many analogies between the political institutions of France and England; but then the destinies of the two peoples parted, and became ever more unlike with the passage of time. They resembled two lines which, starting from neighbouring points but at a slightly different angle, the longer they become, the more indefinitely fall apart.[97]

Hence by the seventeenth century there was a great difference. All over the Continent, there was 'caste', that is to say a system of stratification based on legal differences between groups arising from blood and birth and re-inforced by marriage rules, and its accompaniment, political absolutism.

> As all European monarchies became absolute about the same time, it is not probable that the constitutional change was due to accidental circumstances which occurred simultaneously in every country. The natural supposition is that the general change was the fruit of a general cause operating on every country at the same moment.[98]

This he partly relates, as had Montesquieu, to the reception of Roman law, for in its principles 'to do with the relations between subjects and sovereign... it is full of the spirit of the age when the last additions were made to its compilation – the spirit of slavery'.[99] Tocqueville's summary of the process is given in a footnote to *Ancien Régime*.

At the close of the Middle Ages the Roman law became the chief and almost the only study of the German lawyers, most of whom, at this time, were educated abroad at the Italian universities. These lawyers exercised no political power, but it devolved on them to expound and apply the laws. They were unable to abolish the Germanic law, but they did their best to distort it so as to fit the Roman mould. To every German institution that seemed to bear the most distant analogy to Justinian's legislation they applied Roman law. Hence a new spirit and new customs gradually invaded the national legislation, until its original shape was lost, and by the seventeenth century it was almost forgotten. Its place had been usurped by a medley that was Germanic in name, but Roman in fact.[100]

The causes for the adoption of Roman law all over Europe varied but the effects were similar.

These causes do not suffice to explain the simultaneous introduction of Roman law into every Continental country. I think that the singular availability of the Roman law – which was a slave-law – for the purposes of monarchs, who were just then establishing their absolute power upon the ruins of the old liberties of Europe, was the true cause of the phenomenon. The Roman law carried civil society to perfection, but it invariably degraded political society, because it was the work of a highly civilized and thoroughly enslaved people. Kings naturally embraced it with enthusiasm, and established it wherever they could throughout Europe; its interpreters became their Ministers or their chief agents. Lawyers furnished them at need with legal warrant for violating the law. They have often done so since. Monarchs who have trampled the laws have almost always found a lawyer ready to prove the lawfulness of their acts – to establish learnedly that violence was just, and that the oppressed were in the wrong.[101]

Something very different happened in England because Roman law was never 'received' and common law underpinned both the older 'feudal' institutions and what emerged from them. In England,

> Shutting your eyes to the old names and forms, you will find from the seventeenth century the feudal system substantially abolished, classes which overlap, nobility of birth set on one side, aristocracy thrown open, wealth as the source of power, equality before the law, office open to all, liberty of the press, publicity of debate.... Seventeenth-century England was already a quite modern nation, which has merely preserved in its heart, and as it were embalmed, some relics of the Middle Ages.[102]

In this way it diverged dramatically from what happened elsewhere in Europe. That divergence was the culmination of a much older process. 'It is very probable that at the time of the establishment of the feudal system in Europe what has since been called the "nobility" did not immediately form a caste, but was originally composed of all the chief men of the nation and was thus at first only an aristocracy.' Yet, by the Middle Ages, 'the nobility had become a caste, that is to say, its distinctive mark was birth'. This happened everywhere except England. 'Wherever the feudal system established itself on the continent of Europe it ended in caste; in England alone it returned to aristocracy.' This was the great difference, and one which the English seemed to have overlooked.

> I have always been astonished that a fact, which distinguishes England from all modern nations and which can alone explain the peculiarities of its laws, its spirit, and its history, has not attracted still more than it has done the attention of philosophers and statesmen, and that habit has finally made it as it were invisible to the English themselves. The truth has been often half perceived, half described; never, I think, has the vision of the truth been quite full or quite distinct.

What then is the great difference according to Tocqueville?

> It was far less its Parliament, its liberty, its publicity, its jury, which in fact rendered the England of that date so unlike the rest of Europe than a feature still more exclusive and more powerful. England was the only country in which the system of caste had been not changed but effectively destroyed. The nobles and the middle classes in England followed together the same courses of business, entered the same professions, and, what is much more significant, inter-married.[103]

The contrast with France was immense. There the gap between the different orders grew until they were strangers to each other. The nobility were separated from all other orders: 'for the barrier which separated the nobility of France from the other classes, though very easily crossed, was always fixed and visible; striking and odious marks made it easily recognized by him who remained without. Once a man had crossed the barrier he was separated from all those, whom he had just left, by privileges which were to them burdensome and humiliating.'[104] The worst of these was financial. 'Let us take the most odious of all these privileges, that of exemption from taxation; it is easy to see that from the fifteenth century right down to the Revolution this privilege never ceased to grow.'[105] This could be contrasted with England.

> For centuries past no other inequalities of taxation have existed in England than those successively introduced in favour of the necessitous classes. Notice to what different ends different political principles can lead peoples so close. In the eighteenth century it was in England the poor man who enjoyed exemption from taxation.[106]

The growing gap between the bourgeois and the gentry and the nobility in France was related furthermore to the collapse of local government.

> The fact is that, as the government of the lordship became disorganized, as the meetings of the States-General became rarer or ceased altogether, as the general liberties perished dragging with them in their ruin local liberties, the townsman and the gentleman ceased to have contact in public life. They no longer felt the need of approaching and understanding each other. Every day they became more independent and more unknown to each other. In the eighteenth century this revolution was complete; these two men never met except by mere chance in private life. The two classes were not only rivals, they were enemies.'[107] Furthermore this urban middle class was equally separated from the country dwellers and the poor in the towns. 'Now if . . . we consider this middle class, we see something very similar; the middle class was almost as much separated from the common people as the noble was from the middle class.[108]

How and why had this happened? Others had suggested that 'the English nobility has been more prudent, more clever' than others, and hence survived. In fact, Tocqueville notes, 'The truth is that for a long time past properly speaking there has no longer existed a nobility in England, if the word is taken in its old and circumscribed sense that it has everywhere else retained.'[109] As to when the nobility disappeared, 'This singular revolution is lost in the darkness of past ages', but 'there remains still a living witness, namely, idiom. For several centuries past the word "gentleman" has entirely changed its meaning in England, and the word "roturier" no longer exists.' Thus one could follow the changing sense of the word 'gentleman' as an indication of the divergence of the two civilizations: 'you will see its meaning widen in England in proportion as classes draw nearer and mingle with each other. In each century it is applied to men placed ever a little lower in the social scale. With the English it passes finally to America. In America it is used to designate all citizens without distinction.' In fact, 'Its history is that of the democracy itself.' In France, however, 'the word "gentilhomme" has always been strictly confined to its original sense... The word has always been preserved intact to design the members of the caste, because the caste itself has always been preserved, as separate from all the other classes as it has ever been.'[110]

Thus while in England the barriers, as represented by the word gentleman, evaporated, in France 'this caste had become very much more separated than it was at the time when the word originated, and that a movement took place amongst us exactly the opposite of that which took place in England'.[111] On the other hand, in England, the middle classes and aristocracy were overlapping. It is not that the English aristocracy 'was open but rather due to the fact, as has been said, that its form was indistinct and its limit unknown – less because it was possible to enter than because you never knew when you had got there...'[112]

Tocqueville's solution contains a paradox. On the one hand, England could be seen to be the 'most feudal' of countries in that it had maintained the early spirit of the feudalism which had existed up to the twelfth century all over Europe. Thus Tocqueville could write of William the Conqueror, 'in spite of the revolutions which followed, his version of the feudal system is nevertheless by and large the one that caused the least harm and left the smallest legacy of hatred'.[113] On the other hand, it could equally be argued that, as Tocqueville put it, by the seventeenth century the feudal system was 'substantially abolished', and only a few 'relics of the Middle Ages' remained. Likewise, with France one could argue that it was very un-feudal by the seventeenth century, that is to say it had moved towards caste and absolutism, both of which were diametrically opposed to 'feudalism' of the early period. On the other hand, one could look on the whole *ancien régime* fabric as a distorted form of feudalism. Thus Tocqueville believed that the French revolution destroyed a whole pattern of feudalism: 'ancient institutions were still mixed up with it, and, as it were, interlaced with almost all the religions and political laws of Europe, they had further supplied a crowd of ideas, sentiments, habits, manners, which, so to speak, were adhesive to them...'[114]

This paradox was linked to another. When Tocqueville visited England in 1835 he found it overwhelmingly aristocratic. He was taken aback at the huge estates and country houses. When he returned for his last visit in 1857 'I found England more aristocratic in appearance, at least than I left it twenty years ago. The democratic ferment that then had risen to the surface has disappeared, and all the superior classes seem to have reached a better understanding.'[115] He asked his readers to remember that France was now the 'democratic' nation – and indeed had never had an aristocracy of the English kind; 'in England you have an aristocracy and powerful local influences, while we in France have nothing of the sort.'[116]

He explained this further when he explored the difference between *de facto* and *de jure* equality. The fact that inequalities on the basis of birth had been abolished, or never properly arisen in England, did not mean that there was little inequality. Ironically, the aristocracy were flourishing in eighteenth century England while they were decaying all over Europe.

> This gradual impoverishment of the nobles was seen more or less not only in France but in all parts of the continent where the feudal system, as in France, disappeared without being replaced by a new form of aristocracy. Among the German peoples, who bordered the Rhine, this decay was everywhere visible and much noticed. The contrary was only met with in England. In England the old noble families which still existed had not only preserved, but also had largely increased their wealth...[117]

Thus one found in England, 'Apparent equality, real privileges of wealth, greater perhaps than in any country in the world'.[118] Of course they proclaimed the universal rights and equality of men. But what did these consist of? 'The English have left the poor but two rights: that of obeying the same laws as the rich, and that of standing on an equality with them if they can obtain equal wealth.'[119]

This clash between a *de jure* situation where everyone in theory was equal, but some definitely ended up 'more equal than others', to use Orwell's famous capturing of the paradox, was made worse by the loss of religious faith. In many societies, the poor could reconcile themselves to their status by realizing that there was no alternative: they were born into a fixed social position. This was determined by *karma*, by their activities in previous lives. It was not their fault, a result of their fecklessness or inability. Even Christianity had provided the solace that even if this life was one of poverty, there would be recompense in eternity. The rich would find it virtually impossible to get through the eye of the needle into heaven. The poor and meek would inherit the earth, and heaven too. Yet as faith evaporated, mankind was faced not only with physical misery, but no consolation prize in the after life. How were inequalities to be borne 'in an epoch when our view into another world has become dimmer, and the miseries of this world become more visible and seem more intolerable?'[120]

The solace of God-given inequality was no longer available. Indeed some of Tocqueville's most perceptive observations concern the receding rainbow's end of the constant striving for an ephemeral equality.

> Among democratic peoples men easily obtain a certain equality, but they will never get the sort of equality they long for. That is a quality which ever retreats before them without getting quite out of sight, and as it retreats it beckons them on to pursue. Every instant they think they will catch it, and each time it slips through their fingers.[121]

There develops endless competition as people strive to reach beyond others, but only temporarily succeed. 'They have abolished the troublesome privileges of some of their fellows, but they come up against the competition of all. The barrier has changed shape rather than place. When men are more or less equal and are following the same path, it is very difficult for any of them to walk faster and get out beyond the uniform crowd surrounding and hemming them in.'[122]

*

Tocqueville had thus followed a chain of argument. 'America' was the first truly 'modern' civilization. Much of its modernity had, however, been received from an England which was already very 'modern' by the seventeenth century. England's peculiarity in this respect, its divergence from the other Continental powers had occurred in the period between the twelfth and seventeenth centuries. England had proceeded towards a balanced constitution and an open and competitive social structure while much of the Continent, including Tocqueville's France, had moved toward political centralization and an increasingly rigid stratification based on birth differences. He had thus put forward a thesis to explain how the modern world had emerged. Yet he still needed further answers as to why England was different.

In England the middling-level institutions were retained. In France and most countries, the 'dissolution of the State' phase of early feudalism then coagulated into an absolutism where there were no powerful counter-powers to stop the monopoly power of the monarchy. He found it difficult to explain why the difference originally occurred, or how the English managed to preserve the balance between anarchy and absolutism. Sometimes he put it down to chance. 'Lucky difficulties which obstruct centralisation in England; laws, habits, manners, English spirit rebellious against general or uniform ideas, but fond of peculiarities. Stay-at-home tastes introduced into political life.'[123] At other times he explains it by peculiarly modest and good-natured ruling powers. 'I admit, however, that in order to enable a government in which the supreme power is divided to be permanent to last, as yours has done, for centuries, the ruling authorities must possess an amount of patience and forbearance which never has been granted to ours.'[124] Neither of these appear to be very convincing and probably his most convincing guess, again following Montesquieu, lay in the nature of the effects of islandhood on the nature of warfare and hence on the chances of liberty.

Tocqueville was not a pacifist. He wrote 'I do not wish to speak ill of war; war almost always widens a nation's mental horizons and raises its heart.'[125] On the other hand, his visit to America made him convinced that the absence of powerful, warlike neighbours was an important and necessary, if not sufficient, cause of liberty. He asked the question 'How, then, does it come about that the

American Union, protected though it be by the comparative perfection of its laws, does not dissolve in the midst of a great war? The reason is that it has no great wars to fear.'[126] This was because 'The American Union has no enemies to fight. It is as solitary amid the wilderness as an island in the ocean.'[127] He noted that 'From Canada to the Gulf of Mexico there are only some half-destroyed savage tribes, which six thousand soldiers drive before them.'[128] The New World was 'Placed in the middle of a huge continent with limitless room for the expansion of human endeavour, the Union is almost as isolated from the world as if it were surrounded on all sides by the ocean.'[129] Thus 'The great good fortune of the United States is not to have found a federal Constitution enabling them to conduct great wars, but to be so situated that there is nothing for them to fear.'[130] He was enormously impressed. 'How wonderful is the position of the New World, where man has as yet no enemies but himself. To be happy and to be free, it is enough to will it to be so.'[131]

Tocqueville was not so naive as to think that absence of powerful enemies was a necessary and sufficient explanation for liberty. As he pointed out 'geography gave the Spaniards of South America equal isolation, and that isolation has not prevented them from maintaining great armies. They have made war on one another when there were no foreigners to fight. It is only the Anglo-American democracy which has so far been able to maintain itself in peace.'[132] Thus one needed a combination of a 'point of departure' of liberty and absence of standing armies, combined with no warlike neighbours. 'Their fathers gave them a love of equality and liberty, but it was God who, by handing a limitless continent over to them, gave them the means of long remaining equal and free.'[133]

The situation in continental Europe, as Tocqueville who was brought up in the later years of Napoleon knew only too well, was very different. 'Apart from our continental position, which has always made us feel more acutely the need for concentration of power, decentralisation has never appeared to us as anything but a breakup of the essential rights of sovereignty, that is to say, as the most oppressively active agent of anarchy.'[134] What tended to happen, as Montesquieu had pointed out, was that a powerful nation was sucked into military aggrandisement, or defensive measures, and this almost inevitably led to increased armies, taxation, bureaucracy and absolutism.

> For that reason all nations that have had to engage in great wars have been led, almost in spite of themselves, to increase the powers of the government. Those which have not succeeded in this have been conquered. A long war almost always faces nations with this sad choice: either defeat will lead them to destruction or victory will bring them to despotism.[135]

He noted that 'it is chiefly in time of war that people wish, and often need, to increase the prerogatives of the central government'.[136] Or again, 'All those who seek to destroy the freedom of the democratic nations must know that war is the surest and shortest means to accomplish this.'[137] It was a vicious circle, where success in war was as dangerous as defeat. 'All men of military genius are fond of centralisation, which increases their strength; and all men of centralising genius are fond of war...'[138] Thus, as Montesquieu had shown, 'Any long war always entails great hazards to liberty in a democracy. Not that one needs apprehend that after every victory the conquering generals will seize sovereign power by force after the manner of Sulla and Caesar.'[139]

There was another difficulty, also anticipated by Montesquieu. A small but successful nation or city state would in time be gobbled up by powerful neighbours. As Tocqueville puts it, 'If for a century a democratic country were to remain under a republican government, one can believe that at the end of that time it would be richer, more populated, and more prosperous than neighbouring despotic states; but during that century it would often have run the risk of being conquered by them.'[140] 'As a result of this, except in peculiar circumstances, small nations always end up by being forcibly united with great ones by combining among themselves.'[141] Yet it was these very smaller nations – Greece, the Italian city states, Switzerland, Holland, which were the birth place of liberty. 'Hence at all times small nations have been the cradle of political liberty. It has happened that most of them have lost this liberty in growing larger, a fact which clearly shows that their freedom was more of a consequence of their small size than of the character of the people.'[142] If they decide to counteract the dangers by expanding, as did Rome, the burden of empire was likely to lead to the same dangers. As Montesquieu had argued in different words, 'All passions fatal to a republic grow with the increase of its territory, but the virtues which should support it do not grow at the same rate.'[143] Tocqueville could conclude that in general 'nothing is more inimical to human prosperity and freedom than great empires.'[144]

There seemed no way round the problem.

> War does not always give democratic societies over to military government, but it must invariably and immeasurably increase the powers of civil government; it must always automatically concentrate the direction of all men and the control of all things in the hands of government. If that does not lead to despotism by sudden violence, it leads men gently in that direction by their habits.[145]

In order to survive, a country was pushed towards disaster, '... the great ones prosper not because they are large but because they are strong. Therefore force is often for nations one of the primary conditions of happiness and even of existence.'[146] Unfortunately 'Reason suggests and experience proves that there is no lasting commercial greatness unless it can, at need, combine with military power.'[147]

It was with these difficulties in mind that Tocqueville was particularly impressed, in different ways, by England and the United States. England seemed to be free and to maintain a huge empire – but then she was an island. America was a vast nation, peaceable and free, and, as Tocqueville put it, America was 'as solitary... as an island in the ocean'. How could other, continental, nations break out of the trap? Tocqueville's only solution seems to have been along the lines developed by Montesquieu and Smith, namely that growing trade would finally make international warfare a disaster.

> As the spread of equality, taking place in several countries at once, simultaneously draws the inhabitants into trade and industry, not only do their tastes come to be alike, but their interests become so mixed and entangled that no nation can inflict on others ills which will not fall back on its own head. So that in the end all come to think of war as a calamity almost as severe for the conqueror as for the conquered.[148]

This was his hope, for every revolution and war tended to tip the balance against his precious liberty.

12
Liberty, Wealth and Equality

The compelling feature of Tocqueville's analysis is that he captures the basic contradictions within the new commercial, democratic system that was only half apparent in England but clear in America. He saw that the new system created growing short-term inequalities of wealth, yet this was necessary for it to work. In a variant of Mandeville, he wrote: 'inequality itself will work to forward the wealth of all, for, everybody hoping to come to share the privileges of the few, there would be a universal effort, an eagerness of all minds directed to the acquisition of well-being and wealth.'[1] He saw that the acquisitive spirit was one of the motors for growth: 'an immoderate desire to grow rich, and to do so rapidly; perpetual instability of purpose, and a continual longing for change; a total absence of established customs and traditions; a trading and manufacturing spirit which is carried into everything, even where it is least appropriate.'[2] He saw the strength of the new technologies, but he also saw the future ecological destruction.

He admired the optimism and progressiveness of his American hosts, their 'belief in the wisdom and good sense of mankind; the perfectibility of the human race is contradicted by few, if any'.[3] Yet his own experience and that of his parents showed that this Rousseauite or Godwinian utopianism was a delusion. The best one could do was to choose between evils, as in his advice in relation to France. It was no longer possible to return to the old, aristocratic, world. The revolution had happened and so

> the only choice lay between two inevitable evils; that the question had ceased to be whether they would have an aristocracy or a democracy, and now lay between a democracy without poetry

or elevation indeed, but with order and morality; and an undisciplined and depraved democracy, subject to sudden frenzies, or to a yoke heavier than any that has galled mankind since the fall of the Roman Empire.[4]

This is why it is impossible to characterize Tocqueville as either optimist or pessimist. Like all our thinkers, he showed a little, temporary, optimism, yet at heart he realized that in every success there simultaneously lay a failure, in every step of progress there was a loss. Hope and despair were mixed in about equal proportions. Liberty, equality and wealth might now be irreversible in England and America, but each of them also debased and isolated men.

Tocqueville was fully aware of the negative effects of the peculiar commercial and manufacturing developments in England and America. One was a human cost during the growing industrial and capitalist process which was pitifully obvious half a century on. There was the increasing inequality of wealth generated by machinery replacing human labour, a theme later taken up by Marx as one of the principal reasons for the inevitable collapse of capitalism. Tocqueville noted in Manchester that 'In this factory wages have a tendency to go down. Labour-saving devices are constantly being invented and, by increasing the competition among the workers, bring down the level of wages.'[5] He saw the destitution of workers, in particular the migrant Irish in slums in the midland and northern cities and wrote, 'Here humanity attains its most complete development and its most brutish; here civilization works its miracles, and civilized man is turned back almost into a savage.'[6] There was a contradiction between increasing efficiency and increasing inhumanity, as Adam Smith had realized.

*

Yet the subject which obsessed Tocqueville above all others was the threat to individual liberty posed by the new form of civilization which he saw revealed in America. From his family's experience during the Revolution, and from his political experience during the various upheavals in France, he was well aware of the danger. Like his mentor, Montesquieu, he was terrified of the tendency towards absolutism and political repression. He believed that eternal vigilance was the price of freedom; 'to live in freedom one must grow used to a life full of agitation, change and danger; to keep alert the whole time with a restless eye on everything around; that is the price of freedom.'[7] The difficulty was that political freedom consisted of walking a tightrope. Monarchical governments, as Montesquieu had shown, tended towards absolutism. The history of continental Europe had shown that.[8] What is new about Tocqueville's thought is that with the experience of America he could see that the supposed antidote to this, democracy, was just as dangerous.

Tocqueville's awareness of the fragility of liberty and his pessimism is shown throughout his life. He believed that 'To be free one must be able to invent and persevere in a difficult enterprise, to be able to act on one's own; to live free, one must become accustomed to an existence full of agitation, movement and peril...'[9] For 'political liberty is easily lost; neglect to hold it fast, and it is gone.'[10] 'For my part, I owe that I have no confidence in the spirit of liberty which seems to animate my contemporaries.'[11] He believed that there was a natural tendency towards political absolutism which lay embedded in the drive towards democracy itself. The tendency was not in doubt. 'Reflecting on what has already been said, one is both startled and alarmed to see how everything in Europe seems to tend toward the indefinite extension of the prerogatives of the central power and to make the status of the individual weaker, more subordinate, and more precarious.'[12] Anyone observing current affairs 'will see that in the last half century centralization has increased everywhere in a thousand different ways. Wars, revolutions, and conquests have aided its advance...'[13] Hence 'the social power is constantly increasing its prerogatives; it is becoming more centralized, more enterprising, more absolute, and more widespread.'[14]

The State is a predatory institution which sucks more and more power to itself. 'Thus the state is by no means satisfied by attracting all business to itself, but is more and more successful in deciding everything by itself, without control and without appeal.'[15] It almost automatically increases in power. 'Society, which is in full progress of development, constantly gives birth to new needs, and each one of them is for government a new source of power; for it alone is in a position to satisfy them.'[16] Thus the tendency towards increasing centralization and absolutism did not need a conscious plan on the part of would-be dictators. As he noted of the centralization in France,

> There is nothing to show that, to achieve this difficult result, the government of the "old order" followed a plan carefully thought out before hand; it only gave free play to the instinct, which leads every government to wish for the exclusive management of everything, an instinct which remained always the same despite the diversity of its agents.[17]

The danger is all the greater because the process is simple and almost invisible. 'If the lights that guide us ever go out, they will fade little by little, as if of their own accord.'[18] Despotism is the easy path. 'Thus the art of despotism, once so complicated, has been simplified; one may almost say that it has been reduced to a single principle.'[19] Freedom is hard, despotism easy.

> It cannot be repeated too often: nothing is more fertile in marvels than the art of being free, but nothing is harder than freedom's apprenticeship. The same is not true of despotism. Despotism often presents itself as the repairer of all the ills suffered, the support of just rights, defender of the oppressed, and founder of order. Peoples are lulled to sleep by the temporary prosperity it engenders, and when they do wake up, they are wretched. But liberty is generally born in stormy weather, growing with difficulty amid civil discords, and only when it is already old does one see the blessings it has brought.[20]

What Tocqueville foresaw, in fact, was a new kind of bureaucratic despotism, based on mind-numbing routines rather than brute force and fear.

> Having thus taken each citizen in turn in its powerful grasp and shaped him to its will, government then extends its embrace to include the whole of society. It covers the whole of social life with a network of petty, complicated rules that are both minute and uniform, through which even men of the greatest originality and the most vigorous temperament cannot force their heads above the crowd.[21]

Having seen the dangers, Tocqueville dedicated much of his life to opposing this tendency. 'To explain to men how to escape tyranny, that is the idea of both my books.'[22] His urge to do so arose from two sources. Firstly he loved liberty above everything else.

Like Montesquieu, he saw it as more important than wealth, equality or anything else. Near the end of the second *America* he wrote: 'I think that at all times I should have loved freedom, but in the times in which we live, I am disposed to worship it.'[23] He loved it because of what it did for individuals and for the nation. 'For me, it is self-evident that liberty is the necessary condition, without which there has never been a truly great and virile nation.'[24] Liberty of the individual from governmental control leads to

> the ripening of individual strength which never fails to follow therefrom. Each man learns to think and to act for himself without counting on the support of any outside power which, however watchful it be, can never answer all the needs of man in society. The man thus used to seeking his well-being by his own efforts alone stands the higher in his own esteem as well as in that of others.[25]

On the contrary bureaucratic absolutism led to the crushing of individual responsibility and imagination, and ultimately set the citizen at odds with the state machine.

Tocqueville summarized his deep attachment to liberty in the following moving passage.

> That which in all ages has so strongly attached to it the hearts of certain men as its own attractions, its own charm, quite apart from any material advantages; it is the joy of being able to speak, to act, to breathe, without restraint under no sovereign but God and the law. He who desires in liberty any thing other than itself is born to be a servant. Certain nations pursue it obstinately through all kinds of peril and misfortune. It is not for the material blessings, which it brings, that they love it; they regard liberty itself as a blessing so precious and so necessary, that no other good could console them for its loss, and with its enjoyment they console themselves for the loss of everything else. Others grow weary of it in the midst of their material prosperity; they let it be snatched from their hands without resistance in fear of risking by an effort the very well-being, which they owe to it. What is wanting to those last to remain free? Why? The very desire for freedom.[26]

Tocqueville realized that while liberty also brought long term benefits, in the shorter term one might have to choose between liberty and other desirable things. The true love of liberty pursued it as an end, and not as a means.

> I no longer think that the true love of liberty is ever born from the mere view of the material comforts that it secures; for this view is often darkened. It is very true that in the long run, liberty always brings to those who know how to retain it, ease, comfort, and often riches; but there are occasions, when for the time being, it disturbs the enjoyment of these blessings; there are other occasions, in which despotism alone can give the transient enjoyment of them. Men who only prize liberty for these blessings have never long preserved it.[27]

Tocqueville's passionate love of liberty would have been useless if he had felt that the situation was hopeless, the tendency to absolutism an inevitable progression. In fact he had some hope. In a letter of 1831 he wrote:

> I avow that nonetheless I still hope more than I fear. It seems to me that in the midst of our chaos I perceive one incontestable fact. This is that for forty years we have made immense progress in the practical understanding of the ideas of liberty. Nations, like private people, need to acquire an education before they know how to behave. That our education advances, I cannot doubt.[28]

Towards the end of the second *America* he explained: 'I have sought to expose the perils with which equality threatens human freedom because I firmly believe that those dangers are both the most formidable and the least foreseen of those which the future has in store. But I do not think that they are insurmountable.'[29] He believed that 'Providence did not make mankind entirely free or completely enslaved. Providence has, in truth, drawn a predestined circle around each man beyond which he cannot pass; but within those vast limits man is strong and free, and so are peoples.'[30]

Fifteen months before his death, Tocqueville summarized his hopes and beliefs in a letter to the racist thinker Gobineau.

> To me, human societies, like persons, become something worth while only through their use of liberty. I have always said that it is more difficult to stabilize and to maintain liberty in our

new democratic societies than in certain aristocratic societies of the past. But I shall never dare to think it impossible. And I pray to God lest he inspire me with the idea that one might as well despair of trying. No, I shall not believe that this human race, which is at the head of all visible creation, has become that bastardized flock of sheep which you say it is, and that nothing remains but to deliver it without future and without hope to a small number of shepherds who, after all, are not better animals than are we, the human sheep, and who indeed are often worse.'[31]

*

What then could he do to help to avoid the growing dangers? The first step was to show that the very force which many people thought was delivering mankind from old style despotism contained within itself a tendency towards an even greater and more powerful tyranny. Tocqueville saw that, as part of that inevitable tendency towards equality of opportunity, there would also be an inevitable tendency towards some sort of political participation or 'democracy', rule by the people. Thus he wrote 'The century is primarily democratic. Democracy is like a rising tide; it only recoils to come back with greater force, and soon one sees that for all its fluctuations it is always gaining ground. The immediate future of European society is completely democratic; this can in no way be doubted.'[32] Yet this merely filled him with apprehension. Writing of America he warned that 'This effect of democracy, joined to the extreme instability, the entire absence of coherence or permanence that one sees here, convinces me every day more and more, that the best government is not that in which all have share, but that which is directed by the class of the highest moral principle and intellectual cultivation.'[33] He believed that 'The realistic doctrine carried into politics leads to all the excesses of democracy; it facilitates despotism, centralization, contempt for individual rights, the doctrine of necessity'.[34] 'I therefore think that despotism is particularly to be feared in ages of democracy.'[35] For 'I am convinced that no nations are more liable to fall under the yoke of administrative centralization than those with a democratic social condition.'[36]

In a draft of a letter he summarized the message of the first part of *Democracy in America* as follows.

I had become aware that, in our time, the new social state that had produced and is still producing very great benefits was, however, giving birth to a number of quite dangerous tendencies. These seeds, if left to grow unchecked, would produce, it seemed to me, a steady lowering of the intellectual level of society with no conceivable limit, and this would bring in its train the mores of materialism and, finally, universal slavery. I thought I saw that mankind was moving in this direction, and I viewed the prospect with terror... My aim in writing [my] book was to point out these dreadful downward paths... to make these tendencies feared by painting them in vivid colours... to teach democracy to know itself, and thereby to direct itself and contain itself.[37]

Thus 'To show men if possible how in a democracy they may avoid submitting to tyranny, or sinking into imbecility, is the theme of my book...'[38]

*

One of Tocqueville's great achievements was to see the way in which two planes which were normally held apart, the vertical one of social stratification, and the horizontal one of inter-personal relations, were actually part of the same thing. He realized that the changes he saw from a basically status- (birth-) based society to a contractual (achievement) one had immense effects on social relations. His basic insight was that there was a tension, inconsistency, mutual exclusion between two of the great themes of the French revolution, namely equality and fraternity. The essence of the problem was that 'Equality puts men side by side, without a common link to hold them firm.'[39] Instead of being links in a chain between past and future, or members of a group, they were 'free', but totally isolated individuals. Thus the danger of the new world that was emerging was that 'Men being no longer attached to one another by any tie of caste, of class, of corporation, of family, are only too much inclined to be preoccupied only with their private interests... to retire into a narrow individualism.'[40] This was the new form of individualism which had been proclaimed in eighteenth century Enlightenment philosophy, in the work of Montesquieu, Smith or the other French philosophers. It was a world of 'no grades in society, no classes distinct, no fixed ranks; a people composed of individuals almost alike and wholly equal.'[41]

He believed that this was a relatively recent phenomenon, certainly in France. 'Our ancestors had not got the word "Individualism" – a word which we have coined for our own use, because in fact in their time there was no individual who did not belong to a group, no one who could look on himself as absolutely alone.'[42] French society in the past had been based on exclusive and inclusive groupings, separate and antagonistic. Thus 'each of the thousand little groups, of which French society was composed, thought only of itself'.[43] His distinction between the older form of group 'selfishness', and the new individualism is put in the following passage.

> 'Individualism' is a word recently coined to express a new idea. Our fathers only knew about egoism. Egoism is a passionate and exaggerated love of self which leads a man to think of all things in terms of himself and to prefer himself to all. Individualism is a calm and considered feeling which disposes each citizen to isolate himself from the mass of his fellows and withdraw into the circle of family and friends; with this little society formed to his taste, he gladly leaves the greater society to look after itself. Egoism springs from a blind instinct; individualism is based on misguided judgment rather than depraved feeling. It is due more to inadequate understanding than to perversity of heart.[44]

Thus Tocqueville was very keen to distinguish 'individualism', which saw the person as part of a set of mutual responsibilities, from egoism, which was pure selfishness. He put this in terms of an aphorism. 'So wrong is it to confound independence with liberty. There is nothing less independent than a free citizen.'[45] If the citizen became too independent and egotistic, he would stop being a citizen.

> If the citizens continue to shut themselves up more and more narrowly in the little circle of petty domestic interests and keep themselves constantly busy therein, there is a danger that they may in the end become practically out of reach of those great and powerful public emotions which do indeed perturb peoples but which also make them grow and refresh them.[46]

On the other hand, citizens should have some personal free space. 'From this derives the maxim that the individual is the best and only judge of his own interest and that society has no right to direct his behaviour unless it feels harmed by him or unless it needs his concurrence.'[47] It was a difficult balance and one which he thought the Americans were more successful in achieving than his French contemporaries. 'Every American has the sense to sacrifice some of his private interests to save the rest. We want to keep, and often lose, the lot.'[48]

What Tocqueville thought was that the growing equality would lead to a surfeit of egoism. This would be disastrous politically, but it would also have other undesirable effects. For instance, as we have seen, it altered man's sense of history, making him present-centred, e-historical.[49] Thus, especially in America, the roots were cut off and society was constantly being reinvented. It was not just that it was a new country, but the social structure led people to start again in each generation. Secondly, it led directly into that *Lonely Crowd* which David Riesman, one of Tocqueville's greatest disciples, analysed so well. 'Thus, not only does democracy make men forget their ancestors, but also clouds their view of their descendants and isolates them from their contemporaries. Each man is for ever thrown back on himself alone, and there is danger that he may be shut up in the solitude of his own heart.'[50] The loss to humanity would be immense. 'I fear that the mind may keep folding itself up in a narrower compass for ever without producing new ideas, that men will wear themselves out in trivial, lonely, futile activity, and that for all its constant agitation humanity will make no advance.'[51] Yet it would only be a temporary state, for in the weakness of atomized individuals there would be a tendency for the power of the State to increase. 'As the extent of political society expands, one must expect the sphere of private life to contract.'[52]

The real problem was 'How to reconcile equality, which separates and isolates men, with liberty? How to prevent a power, the offspring of democracy, from becoming absolute and tyrannical? Where to find a force able to contend against this power among a set of men, all equal, it is true, but all equally weak and impotent?'[53] The danger was that, since all power tends to corrupt, there would be a drift towards centralization and hence towards despotism. Tocqueville had seen this happen in France in relation to bureaucratic centralization:

> a taste for holding office and a desire to live on the public money is not with us a disease restricted to either party, but the great, chronic ailment of the whole nation; the result of the democratic constitution of our society and of the excessive centralization

of our Government; the secret malady which has undermined all former governments, and which will undermine all governments to come.[54]

The danger was aggravated by the passions and desires of men.

In a marvellous passage Tocqueville lays out the tendency towards benevolent despotism implicit in American civilization.

> I am trying to imagine under what novel features despotism may appear in the world. In the first place, I see an innumerable multitude of men, alike and equal, constantly circling about in pursuit of the petty and banal pleasures with which they glut their souls. Each one of them, withdrawn into himself, is almost unaware of the fate of the rest. Mankind, for him, consists in his children and his personal friends. As for the rest of his fellow citizens, they are near enough, but he does not notice them. He touches them but feels nothing. He exists in and for himself, and though he still may have a family, one can at least say that he has not got a fatherland. Over this kind of men stands an immense, protective power which is alone responsible for securing their enjoyment and watching over their fate. That power is absolute, thoughtful of detail, orderly, provident, and gentle. It would resemble parental authority if, fatherlike, it tried to prepare its charges for a man's life, but on the contrary, it only tries to keep them in perpetual childhood. It likes to see the citizens enjoy themselves, provided that they think of nothing but enjoyment. It gladly works for their happiness but wants to be sole agent and judge of it. It provides for their security, foresees and supplies their necessities, facilitates their pleasures, manages their principal concerns, directs their industry, makes rules for their testaments, and divides their inheritances. Why should it not entirely relieve them from the trouble of thinking and all the cares of living?[55]

This portrait puts flesh on his idea that 'the type of oppression which threatens democracies is different from anything there has ever been in the world before'.[56] The difference between the despotism of the old tyrannies and the new bureaucratic State was that, 'Under the absolute government of a single man, despotism, to reach the soul, clumsily struck at the body, and the soul, escaping from such blows, rose gloriously above it; but in democratic republics that is not at all how tyranny behaves.'[57] A further contrast lay in the new material affluence which was in itself a product of liberty. As Boesche points out, the

> very prosperity that accompanied bourgeois society might, in Tocqueville's opinion, give birth to the conditions that make this new despotism possible, like a plant whose flowering moment also signals its demise. 'One must take care,' wrote Tocqueville, 'not to confuse political liberty with certain effects it sometimes produces.' Political liberty leads to prosperity, but prosperity leads to 'the taste for material well-being' and to a 'passion for making fortunes'; these in turn threaten to 'extinguish' the very political liberty that gave them birth.[58]

> The men of the eighteenth century hardly knew that kind of passion for material comfort, which is, so to speak, the mother of servitude, an enervating but tenacious and unalterable passion, which readily mingles with and twines itself round many private virtues such as love of family, respectability of life, regard for religious beliefs, and even the assiduous if lukewarm practice of the established worship, which is partial to respectability but forbids heroism, which excels in making men steady but citizens mean-spirited. The men of the eighteenth century were both better and worse.[59]

His deepest worry was that the growing equality and individualism put people in a particularly weak position to stand up to the State. The practice of divide and rule had been a conscious tactic in the old order.

> Almost all the vices, almost all the mistakes, almost all the fatal prejudices which I have just described owed, in fact, either their birth, or their continuance, or their development, to the practice pursued by most of our kings in dividing men in order to govern them more absolutely.[60]

Yet in the new order, such division between individuals became institutionalized. Thus

> when the citizens are all more or less equal, it becomes difficult to defend their freedom from the encroachments of power. No

one among them being any longer strong enough to struggle alone with success, only the combination of the forces of all is able to guarantee liberty. But such a combination is not always forthcoming.[61]

Thus he reported: 'What I find most repulsive in America is not the extreme freedom reigning there but the shortage of guarantees against tyranny.'[62] He saw that there could very easily be a switch from the 'sovereignty of the people' to the sovereignty of the State.

So, for a people who have reached the Anglo-Americans' social state, it is hard to see any middle course between the sovereignty of all and the absolute power of one man... the social state I have just described may lead as easily to the one as to the other of those results.[63]

He saw that Montesquieu's earlier warnings might apply here. 'Montesquieu has noted that nothing is more absolute than the authority of a prince who immediately succeeds a republic, since the undefined powers that had been fearlessly entrusted to an elected magistrate then pass into the hands of a hereditary sovereign. This is true in general but applies more particularly to a democratic republic.'[64]

Tocqueville's solution was, as with equality, to suggest a balance. Too little equality was as bad as too much. The balance must be between too much centralization and too little. He put the continuum clearly as follows.

There are two great drawbacks to avoid in organizing a country. Either the whole strength of social organization is centred on one point, or it is spread over the country. Either alternative has its advantages and its drawbacks. If all is tied into one bundle, and the bundle gets undone, everything falls apart and there is no nation left. Where power is dispersed, action is clearly hindered, but there is strength everywhere.[65]

This idea of a balance became his central concern. As he recalled 'I had conceived the idea of a balanced, regulated liberty, held in check by religion, custom and law; the attractions of this liberty had touched me; it had become the passion of my life...'[66]

This balance reminds one very much of Montesquieu's solution of the balance of the contending forces of law, religion and other institutions. The judicial power was very important as a check to the administration. 'The necessity of bringing the judicial power into the administration is one of those *central* ideas to which I am brought back in all my researches to discover what allows and can allow men the enjoyment of political liberty.'[67] Likewise, the balance between the secular and the religious was also important. Tocqueville warned of the danger of a pact, when religion and politics entered into a union which crushed all liberty. He noted that at the time of the rise of absolutist monarchies in Europe 'the Catholic clergy throughout Europe had become both a religious and a political body.'[68] He warned of a dangerous slavery 'where the Church is so thoroughly in the hands of the State as to become an instrument of government; of this Russia is an example.'[69] The danger had, as Montesquieu knew, been manifest in France. 'The Church of France, under Louis XIV, was both a political and a religious institution.'[70]

*

Tocqueville's central obsession was with the balance between the centre and the periphery. In illustrating his important argument here he drew above all, as Montesquieu had done, on the English case. America was too new and decentralized to provide a case study. The Continental states had clearly fallen off the tightrope. The problem was how to 'unite liberty to the already existing equality', he 'searched eagerly in a democratic country for the fundamental conditions of liberty.'[71] He found these conditions in England. He believed that it had managed to walk the narrow path between too much and too little centralization, with only a few false steps, for 1000 years.

Tocqueville wrote a summary of the situation in 1835. 'There is a great deal of centralization in England; but of what sort?' To this he answered, 'Legislative and not administrative; governmental rather than administrative.' 'The mania for *regimentation* ... is found here as elsewhere', but unlike in France, it had little effect. This is 'because the *centralizing* power is in the hands of the legislature, not of the *executive*'. Among the 'Lucky consequences' of this were the following:

Publicity, respect for rights, obligation to refer to local authorities for the execution of the law; natural tendency to divide administrative authority so as not to create too strong a rival power. Centralisation very incomplete since it is carried out by a legislative body; *principles* rather than *facts*; *general* in spite of a wish to be *detailed*.[72]

The '*Greatness and strength* of England' was 'explained by the power of centralization in certain matters'. On the other hand, the '*Prosperity, wealth, liberty* of England' were 'explained by its *weakness* in a thousand others'.[73] This mixture was even shown in relation to the Indian Empire. England was

> the most powerful in some things, and the weakest and most embarrassed in some other; which keeps eighty million people under its obedience three thousand leagues away, and does not know how to get out of the smallest administrative difficulties; which excels at taking advantage of the present, but does not know how to foresee the future. Who can find a word to explain all these anomalies?[74]

What Tocqueville noticed was a productive tension between different forces. 'Principle of *centralisation* and principle of *election of local authorities*: principles in direct opposition.' He believed that these were the 'only means of combining the two principles to some extent since the one is essential to the power and existence of the State, the second to its prosperity and liberty'. This was the key. 'England has found no other secret', and France must learn it. 'The whole future of free institutions in France depends on the application of these same ideas to the genius of our laws.' If one could find a way 'to subject the centralizing power to publicity' and 'to have its *local* decisions carried out by *elected* authorities', Tocqueville would see 'no objection to extend its power as much as you like....'[75]

He described this balance on several other occasions. The 'English government is strong although the localities are independent'.[76] He quoted Dr. Bouring to the effect that 'England is the country of decentralization. We have got a government, but we have not got a central administration. Each county, each town, each parish looks after its own interests.'[77] In comparing France to England, Tocqueville wrote in 1853, in England you have an aristocracy and powerful local influences, while we in France have nothing of the sort. You have no centralization, while we have centralized the administration more than perhaps has ever been done in a great country. Whence it results that in England corruption and intimidation are the instruments chiefly of the great landowners, and of the rich in general, while with us corruption and intimidation can be made use of only by the Government.[78]

The heart of the difference lay in the fact that the English had centralized the judicial but not the administrative system. 'The English are the first people who ever thought of centralizing the administration of justice. This innovation, which dates from the Norman period, should be reckoned one of the reasons for the quicker progress which this nation has made in civilization and liberty.'[79] In France, the early divisive tendencies of feudalism went in the other direction. The barons became too powerful. 'That is what happened in France, where the barons went so far as to abolish the right of appeal to the king's courts. That is what did not happen in England. William, master of all, gave lavishly but kept still more.'[80] Ironically, Tocqueville's Norman predecessor, William the Conqueror, managed to steer a middle course. Faced with too much or too little centralization, Tocqueville wrote, 'I don't know if a mean between these extremes can be found, but it would seem that William did find it.'[81]

The contrast with his three other cases, America, France and China, was instructive. In America there was as yet an almost complete absence of centralization. 'There is nothing centralized or hierarchic in the constitution of American administrative power, and that is the reason why one is not at all conscious of it. The authority exists, but one does not know where to find its representative.'[82] Thus 'Nothing strikes a European traveller in the United States more than the absence of what we would call government or administration.'[83] That is because 'there is no central point on which the radii of administrative power converge'.[84] The problem lay in the future, for as the country grew wealthier and more populous, there would be a tendency towards bureaucratic centralization.

On the other hand, France and other continental powers represented the other extreme. The height of centralization had been reached in France in the later seventeenth and early eighteenth centuries. 'Under Louis XIV France reached the greatest possible degree of centralization of government that can be conceived, for one man made the general laws and had the power to interpret them, and he represented France abroad and acted in her name. "I am the state," he said, and he was right.'[85] But after the disruption of the Revolution, Napoleon has been quick to start the process again and now 'I assert that there is no country in Europe in which public administration has not become not only more centralized but also more inquisitive and minute.'[86] 'Among all the nations of continental Europe, one may say that there is not one that understands communal liberty. However, the strength of free peoples resides in the local community.'[87] The new socialist movements which were sweeping across Europe provided no alternative to this. As Drescher writes, 'In socialism he saw only the logical culmination of an omnipotent centralizing urge combined with a contempt for man as individual and citizen. It was "a new form of servitude".'[88]

Tocqueville saw China as the extreme of bureaucratic centralization. He noted that China had benefited from long periods of peace and order. 'China ... had existed in peace for centuries; her conquerors had adopted her mores; order prevailed. Material prosperity of a sort was visible everywhere. Revolutions were very rare and war, one might almost say, unknown.'[89] Yet there was the famous stagnation.

> Three hundred years ago, when the first Europeans came to China, they found that almost all the arts had reached a certain degree of improvement, and they were surprised that, having come so far, they had not gone further. Later on they found traces of profound knowledge that had been forgotten. The nation was a hive of industry; the greater part of its scientific methods were still in use, but science itself was dead.[90]

This withering away of curiosity and creativity was very puzzling.

> The Chinese, following in their fathers' steps, had forgotten the reasons which guided them. They still used the formula without asking why. They kept the tool but had no skill to adapt or replace it. So the Chinese were unable to change anything. They had to drop the idea of improvement. They had to copy their ancestors the whole time in everything for fear of straying into impenetrable darkness if they deviated for a moment from their tracks.[91]

Tocqueville's solution to the puzzle was to blame a centralized and uniform bureaucratic system.

> China seems to offer the classic example of the sort of social prosperity with which a very centralised administration can provide a submissive people. Travellers tells us that the Chinese have tranquillity without happiness, industry without progress, stability without strength, and material order without public morality. With them society always gets along fairly well, never very well. I imagine that when China is opened to the Europeans, they will find it the finest model of administrative centralisation in the world.[92]

He touched on a couple of aspects of this system. One was the overwhelming desire for bureaucratic office. 'There is no need for me to say that this universal and uncontrolled desire for official appointments is a great social evil, that it undermines every citizen's sense of independence and spreads a venal and servile temper throughout the nation...'[93] The avenue to such offices was through the examination system. 'In China... no man graduates from one public office to another without passing an examination. He has to face this test at every stage of his career... Lofty ambition can hardly breathe in such an atmosphere.'[94]

It was not that Tocqueville was against government as such. He was not an Anarchist. He believed that strong government and administrative centralization were different things. 'In our day we see one power, England, which has reached a very high degree of centralization of government; there the state seems to move as a single man.'[95] Yet it was a free and wealthy country.

> England, which has done such great things in the last fifty years, has no administrative centralization. For my part, I cannot conceive that a nation can live, much less prosper, without a high degree of centralisation of government. But I think that administrative centralisation only serves to enervate the peoples that submit to it, because it constantly tends to diminish their civic spirit.[96]

Tocqueville also saw the English solution as having another enormous advantage. It made it possible to change peacefully over long periods without needing periodic revolutions. Continuous evolution rather than punctuated equilibria was the advantage of a proper balance between centre and periphery.

*

Like Montesquieu, Tocqueville attempted to elaborate a number of the institutional checks on the tendency towards absolutism. In early notes he quoted an Irish priest who said that 'Freedom of the press, Sir, is the first and perhaps the only efficient weapon which the oppressed has against the oppressor; the weak against the strong; the people against the government and the great.'[97] In relation to America he wrote 'The more I observe the main effects of a free press, the more convinced am I that, in the modern world, freedom of the press is the principal and, so to say, the constitutive element in freedom.'[98] It was particularly important in a democracy. 'The press is, par excellence, the democratic weapon of freedom.'[99] It allowed individuals, weak and fragmented, to coalesce into an imagined community and hence to act as a counter-balance to the State. 'For this reason freedom of the press is infinitely more precious in a democracy than in any other nation.'[100] Thus, as he explained, 'the more equal men become and the more individualism becomes a menace, the more necessary are newspapers. We should underrate their importance if we thought they just guaranteed liberty; they maintain civilization.'[101]

As important as the freedom of the press was the nature of the legal system. As a trained lawyer himself, and a disciple of Montesquieu, Tocqueville was well aware of the power of the law. He saw several features of the Anglo-American system which particularly attracted him. One was the jury system. In his Journal while visiting America he wrote 'The jury is the most direct application of the principle of the sovereignty of the people.'[102] Or as he put it in the finished book: 'Therefore the jury as an institution really puts control of society into the hands of the people or of that class.'[103] He saw the jury as having a double role. 'The jury is both the most effective way of establishing the people's rule and the most efficient way of teaching them how to rule.'[104] In fact it was the second of these that he most strongly commended. 'Juries teach men equity in practice. Each man, when judging his neighbour, thinks that he may be judged himself.'[105] Thus he believed that 'Juries are wonderfully effective in shaping a nation's judgment and increasing its natural lights. That, in my view, is its greatest advantage.

It should be regarded as a free school which is always open and in which each juror learns his rights...'[106]

Another crucial power was the independence of the judiciary, and in particular the institution of justices of the peace. 'The power of the courts has been at all times the securest guarantee which can be provided for individual independence but this is particularly true in ages of democracy.'[107] As for independent magistrates, 'When a justice of the peace has a share in the administration, he brings with him a taste for formalities and for publicity, which renders him a most inconvenient instrument for a despotism; but he is not the slave of those legal superstitions which make magistrates so little capable of administration.'[108] Thus the judiciary should be brought into the administration as much as possible. Independent justices, rather than paid bureaucrats, were essential. 'The necessity of introducing the judicial power into the administration is one of those *central* ideas to which I am led by all my investigations concerning the sources of political liberty.'[109]

Of course there are still dangers. The tendency of the State to grow ever more powerful may mean that it starts to corrupt the judges. 'Thus the government is daily more able to escape the obligation to have its will and its rights sanctioned by another power. Unable to do without judges, it likes at least to choose the judges itself and always to keep them under its hand.'[110] The protection against this is to divide the legislature up into several parts. He wrote of

> the principle of the division of legislative power; henceforth the need to share legislative activity between several bodies has been regarded as a demonstrated truth. This theory, hardly known to the republics of antiquity, introduced into the world almost by chance, like most great truths, and misunderstood by several modern nations, has at last become an axiom of political science in our day.[111]

The best example of this system at work was in New England. 'All the general principles on which modern constitutions rest, principles which most Europeans in the seventeenth century scarcely understood and whose dominance in Great Britain was then far from complete, are recognized and given authority by the laws of New England; the participation of the people in public affairs, the free voting of taxes, the responsibility of government officials, individual freedom, and trial by jury – all these things were established without question and with practical effect.'[112] They had also adopted the other great check on abuse of power, the ability of the people to dismiss the rulers through elections. 'An arbitrary power to dismiss public officials is the only guarantee of that sort of active and enlightened obedience which no judicial sanction can impose. In France we seek the ultimate guarantee in the *administrative hierarchy*; in America *election* fills that role.'[113] All these checks and balances of a formal nature were not, however, enough. Tocqueville devoted much attention to two other areas. One was the necessity for religion, a second was how to mitigate the dangers of individualism through forming associations.

*

Tocqueville's views on religion are surprising, for they contain another paradox. While too much religion, that is religion formally enforced by the State, is disastrous, too little religion is equally dangerous. One might have expected him to advocate a complete separation of politics and religion, but in fact he does not do this. He saw that religion and politics must be combined in some way: 'the real greatness of mankind must arise from the combined action of liberty and religion; the one to animate, the other to restrain.'[114] He particularly admired the way in which this was done in England. Again implicitly echoing Montesquieu's remark about the combination of wealth, liberty and piety, he wrote that 'I enjoyed too, in England what I have long been deprived of – a union between the religious and the political world, between public and private virtue, between Christianity and liberty.'[115] Indeed, he makes the further connection when he writes 'So there must be a hidden relationship between those two words: *liberty* and *trade*. People say that the spirit of trade naturally gives men the spirit of liberty. Montesquieu asserts that somewhere',[116] and further suggests, 'I think it is above all the spirit and habits of liberty which inspire the spirit and habits of trade.'[117] But how did England manage to combine wealth, liberty and religious enthusiasm? How was it that England was so surprisingly active in mixing the latter two, being a country where, for example, 'Great political parties, as always happens in free countries, found their interest in uniting their cause with that of the Church'?[118]

The nearest Tocqueville comes to solving the apparent contradiction is by showing that the English made the separation not between religion and politics, but between the public and the private. Politics belonged to public life, religion to the private. The case was illustrated by English Catholics.

> In fact, I never met with an English Catholic who did not value, as much as any Protestant, the free institutions of his country, or who divided morality into two sections, one consisting of public virtues, which might be safely neglected, and the other of private duties, which alone need be observed.[119]

His insights into the necessary connection between liberty and religion came out of his observations of England and America.

> I have already said enough to put Anglo-American civilisation in its true light. It is the product (and one should continually bear in mind this point of departure) of two perfectly distinct elements which elsewhere have often been at war with one another but which in America it was somehow possible to incorporate into each other, forming a marvellous combination. I mean the *spirit of religion* and the *spirit of freedom*.[120]

He noted: 'One cannot therefore say that in the United States religion influences the laws or political opinions in detail, but it does direct mores, and by regulating domestic life it helps to regulate the state.'[121] Thus he advocated the importance of religion. 'Despotism may be able to do without faith, but freedom cannot.'[122] Or again, 'Society has nothing to fear or hope from another life; what is most important for it is not that all citizens should profess the true religion but that they should profess religion.'[123] As he put it in one of his aphorisms, 'For my part, I doubt whether man can support complete religious independence and entire political liberty at the same time. I am led to think that if he has no faith he must obey, and if he is free he must believe.'[124]

Yet he was also aware from his own Catholic background that there was a tendency in religion to move towards absolutism and indeed be its main support.

> Montesquieu, in attributing a peculiar force to despotism, did it an honour which, I think, it did not deserve. Despotism by itself can maintain nothing durable. When one looks close, one sees that what made absolute governments long prosperous was religion, not fear.[125]

How could the danger of too much religion be avoided? Again it was best if there was a division into balanced and competing units. Following Montesquieu and Smith he took the view that tolerance in religion arose from powerlessness. One religion in a State, for instance Catholicism, would be disastrous. Even if there were two, equally powerful, it would be hopeless.

> If two religions faced each other, we should be cutting each others' throats. But as none has as much as a majority, all need toleration. Besides there is a general belief among us, a belief which I share, that some religion or other is needed by man as a social being.[126]

With its proliferation of sects, in America even the Catholics preached toleration. 'The Catholics are in a minority, and it is important for them that all rights should be respected so that they can be sure to enjoy their own in freedom.'[127] Thus each religious sect was thwarted in its political ambitions.[128]

The result was that in the world of sectarian America or England, the separation between formal religion and formal politics had been effected.

> Religion regards civil liberty as a noble exercise of men's faculties, the world of politics being a sphere intended by the Creator for the free play of intelligence. Religion, being free and powerful within its own sphere and content with the position reserved for it, realises that its sway is all the better established because it relies only on its own powers and rules men's hearts without external support.[129]

Tocqueville had noticed this modesty when he visited England as well. 'I was struck this time in England, as I had previously been, to see how a religious sentiment conserved its power, without becoming something that absorbs and destroys all other motives of human action.'[130] Indeed, he believed that the two were linked. Religious faith was much more active and sincere if it eschewed an alliance with the State. For 'any alliance with any political power whatsoever is bound to be burdensome for religion. It does not need their support in order to live, and in serving them it may die.'[131]

Thus religious faith was needed to unite and animate a democratic peoples, to provide an ideological alternative to the overbearing State and to give ideals and confidence. 'The longer I live the less

I think that the peoples of the world can ever separate themselves from a positive religion...'[132] Yet religious institutions must not become so powerful that they became, as in many ancient despotisms, the most potent force for tyranny.

*

Tocqueville's final major protection against the tendency towards absolutism was his support for associations, or what we might today call a strong 'Civil Society'. Modern society supported the individual, the equality of citizens and the rights of man. Yet in order to effect very much, individuals must cooperate. This led Tocqueville into a discussion of how a modern society which could no longer use birth as the recruiting device to form groups could operate. His answer was that people in such a society generated large numbers of associations instead, that is to say contractual, voluntary, groupings, usually with limited purposes, which would allow individuals to drop some of their narrow egotism and work for a common goal. The importance of such associations was naturally most marked where equality was most extreme, in other words in America and we have seen his treatment of the association in the American context.

The English case puzzled Tocqueville. It appeared to be once again somewhere between the birth-status groups of traditional France, and the individual-associational extreme of America. A contradiction between individual's interest and that of the association seemed to him to be present in England. 'Two spirits which, if not altogether contrary, are at least very diverse, seem to hold equal sway in England.'[133] He could not 'completely understand how the "spirit of association" and the "spirit of exclusion" both came to be so highly developed in the same people, and often to be so intimately combined'.[134] He decided that

> On reflection I incline to the view that the spirit of individuality is the basis of the English character. Association is a means suggested by sense and necessity for getting things unattainable by isolated effort. But the spirit of individuality comes in on every side; it recurs in every aspect of things.[135]

People in England were ultimately individuals, but were prepared to associate as the only means to attain their ends. 'That being so, the need to club together is more generally felt, because the urge to get things is more general and stronger.'[136]

> Example a club; what better example of association than the union of individuals who form the club? What more exclusive than the corporate personality represented by the club? The same applies to almost all civil and political associations, the corporations...[137]

Curiously, therefore, the extreme individualism of the English led to more cooperation between people for specific purposes than the group-mindedness of the French. The absence of any alternative structures 'prompts people to pool their efforts to attain ends which in France we would never think of approaching in this way. There are associations to further science, politics, pleasure, business...'[138] In France, on the other hand, before the Revolution, the country was divided 'into a great number of sections, and within each of these small enclosures there was seen to speak a distinct society, which was only concerned with its own particular interests, and took no part in the life of the whole'.[139] Somehow the Anglo-Saxon peoples, including of course the Dutch, managed to combine individualism and cooperation in an unusual way.

Thus Tocqueville saw the associational forms as having their 'point of departure' in England. 'The English, though the divisions between them are so deep, seldom abuse the right of associations, because they have had long experience of it.'[140] It then spread to America. 'The right of association is of English origin and always existed in America. Use of this right is now an accepted part of customs and of mores.'[141] This was in contrast to the trend on the Continent. In the remote past there had been as many 'associations' in Germany or France as in England. Yet while they had continued and blossomed in England and then America, they had been destroyed on the Continent and their powers absorbed by the increasing power of the Absolutist state. 'The point I want to make is that all these various rights which have been successively wrested in our time from classes, corporations, and individuals have not been used to create new secondary powers on a more democratic basis, but have invariably been concentrated in the hands of the government.'[142] This was disastrous. Like Montesquieu, Tocqueville believed that numerous 'secondary powers', that is associations of free individuals into organizations for running their own affairs, were the major protection against tyranny. Using a metaphor of a dyke used to prevent the flood of despotism he wrote 'In countries where such associations do not exist, if private people did not artificially and temporarily create something like them, I see no other dyke to hold back tyranny of whatever sort, and a great nation might with impunity be oppressed by some tiny faction or by a single man.'[143]

*

The encounter with Tocqueville adds further elements to a possible solution to the riddle we are pursuing. He refines the concept of the separation of powers, the safeguards and importance of liberty, the precarious balance between centre and periphery and the effects of war. Tocqueville saw the key to real progress as a never-ending tension or conflict between institutional spheres and in the absence of a dominating and dominant religion or State. He noted the beneficial effects of commerce on morals, the tendency to predate by war, the importance of an independent judiciary and the power of law, the way in which liberty brought wealth in its train, the way in which America had harmonized self-interest and the public good, the importance of secondary powers and the negative effects of industrialization. All these themes we have encountered in previous thinkers but with him they are given a fresh and deepened treatment.

There are also many new areas that he explored: the importance of the tendency towards 'caste', class and social hierarchy, the effects of growing equality in many spheres, the importance of associations, the separation of public and private. He drew attention to the materialistic ethic of capitalism, the pursuit of profit as an end in itself, the curiously high estimation of work, the effects of commerce on concepts of time, space and the family, the presence of an 'imagined community' as the basis of the modern nation state, the effects of equality on family relations, the dangers of a loss of liberty caused by the rising tide of 'democracy' itself and of centralization, the dangers of egotism and the necessity for religious belief.

Particularly important for our purposes, he supplements Montesquieu and Smith's historical account by giving the most detailed and convincing analysis not only of the difference between England and France, but of how that difference occurred and evolved.

He showed the origins of the American system in medieval and early modern England, the difference between French peasant social structure and English agriculture, the entirely different political history of the two countries, with revolution and rigidity in one and flexible evolution in the other. He noted the absence of a nobility in England and the entirely different meaning of the words 'gentleman' and 'gentilhomme'.

*

We can see that by the time of Tocqueville's death in 1859 the questions concerning the recent development of human civilizations had been clearly posed and a plausible set of hypotheses to answer some of them had been put forward. These answers will probably strike many today as surprisingly different to those with which they are familiar. This is because much of the subsequent work during the century and a half since then has buried both the questions and any possible answers under a heap of alternative approaches so that the earlier work has become increasingly obscured. This inquiry has largely been an excavation to unearth something which was once widely known but is now largely forgotten.

In order to understand both the very great difficulties facing contemporary scholars, and also the continuing vitality of the Enlightenment questions and answers, it is worth briefly considering one last major thinker. He was a man whose work spanned almost all of the second half of the twentieth century and like his Enlightenment predecessors absorbed many of the great traditions of western thought. He was in a certain sense one of the last representatives of the great alternative tradition whose answer to the riddle of how the modern world emerged has been the theme of this inquiry.

IV An Answer to the Riddle?

13
Ernest Gellner and the Conditions of the Exit[1]

Ernest Gellner was born in 1925 in Czechoslovakia, the son of a Jewish journalist turned businessman. The family lived in Prague until the German occupation of 1939, when they moved to England. In 1949 Gellner obtained a first-class degree in philosophy, politics and economics at Oxford. He then went to Edinburgh for two years on an assistantship in philosophy and became a lecturer in Sociology at the London School of Economics. At the LSE he became attracted to anthropology, where Bronislaw Malinowski's influence was still strong. He visited Morocco in 1954, and soon began his fieldwork for an anthropology Ph.D. subsequently published as *Saints of the Atlas* (1969). In 1962 he received a personal chair at the LSE as professor of sociology with special reference to philosophy. He wrote a number of works and collections of essays connecting anthropology, sociology and philosophy. He also continued his studies of Islamic societies, making eight fieldwork visits to Morocco and publishing *Muslim Society* in 1981. Gellner became professor of social anthropology at Cambridge in 1984 and retired in 1993. He died of a heart attack in Prague on 5 November 1995.

Gellner's life had produced a set of contradictions which remind one of our earlier thinkers and help to explain how he revived an interest in the riddle of modernity. Jiri Musil describes the first clash. 'During his childhood in Bohemia... Gellner experienced the last remnants of the old, traditional world, saw the Czech countryside, real villages and farmers... Yet parallel to this, he lived in a dynamic city that in his lifetime became one of the new metropoli of Europe. He could not miss the contrast.' Musil points out that it is this which differentiated him from almost all of his colleagues. 'He did not live only in an urban environment from his early childhood as did most of his later British or American colleagues who studied industrial societies.' Musil believes that 'The others could not be very directly aware of the "metamorphosis", because they were already beyond the great divide.... Gellner's knowledge of Czech agricultural and industrial society ... allowed him to understand better what transpired in European societies from the eighteenth century to the present.'[2]

The early clash between eastern and western Europe in his upbringing was reinforced by at least three further intellectual and social experiences which heightened his awareness of the peculiarities and precariousness of our civilization. One of these was his professional interest in the great philosophical watershed between the *ancien régime* and modernity which took place in the eighteenth century and particularly in the Scotland he had experienced in Edinburgh and read about in his beloved David Hume. Here Gellner found a specification of the foundation of the new world and all its strangeness, which was given further precision by his other mentor, Kant.

The second reinforcement came from his professional involvement with Islam. This provided him with an invaluable countermodel. He approvingly quoted Tocqueville on the fact that 'Islam is the religion which has most completely confounded and intermixed the two powers ... so that all the acts of civil and political life are regulated more or less by religious law'.[3] Islam made Gellner deeply aware that the mixing of religion and politics is the normal state of mankind: their separation in certain parts of the world is a recent peculiarity. The way in which Islam functions despite this lack of separation continued to puzzle him. Islam 'exemplifies a social order which seems to lack much capacity to provide political countervailing institutions or associations, which is atomized without much individualism, and operates effectively without intellectual pluralism'.[4]

Thirdly, there was Gellner's continuing interest in the only other major 'totalitarian' or 'closed' system that existed for most of his lifetime, communism. Whereas Islam embeds politics within religion, the Soviet world tried to embed economy, society and religion within the polity. He wrote that 'Under the Communist system, truth, power and society were intimately fused'.[5] The collapse of this closed world provided Gellner with the chance to undertake a post mortem. The surprise and opportunity perhaps helps to account for the fact that some of his most inspired writing occurred in the last six years of his life, after 1989. As he himself put it,

> It is this collapse which has taught us how better to understand the logic of our situation, the nature of our previously half-felt, half-understood values. We now see the manner in which they emerge from the underlying constraints and strains of our condition. It provides a better way of understanding society and its basic general options.[6]

Thus Ernest Gellner was well placed to see that there is a riddle. In trying to solve it he elaborated and synthesized many of the Enlightenment themes.

*

One characteristic of the emergence of 'modernity' is the growth of rationality or the disenchantment of the world. There is a 'radical discontinuity' which exists 'between primitive and modern mentality'.[7] This is '*the* transition to effective knowledge', which Gellner described many times.[8] This is, of course, not unlike the work of Popper and Kuhn.[9] But Gellner's stress is on the fact that 'The attainment of a rational, non-magical, non-enchanted world is a much more fundamental achievement than the jump from one scientific vision to another'. Popper 'underestimates the difficulties' of establishing an Open Society.[10]

In a number of his earlier works Gellner developed the idea that the separation of cognition or thought was just one example of the most fundamental characteristic of the great transformation, that is the effort to separate and balance the deepest forces in human life – the pursuit of power (politics), wealth (economics), social warmth (kinship) and meaning (religion). Gellner noted that in the majority of human societies, there is no separation of institutions. For instance, in tribal societies there is no distinction between economic and political.[11] But 'Under capitalism, this unity disappears; productive units cease to be political and social ones. Economic activities become autonomous . . .' This separation of the economic from the political and social is one of the important features of western industrial capitalism. 'The really fundamental trait of classical capitalism is that it is a very special kind of order in that the economic and the political seem to be separated, to a greater degree than in any other historically known social form.'[12] He asked how it was that 'Production replaces Predation as the central theme and value of life'.[13]

This separation of spheres, where politics, economics, religion and kinship are artificially held apart, is the central feature of modern civilization. None of the institutions is dominant. There is no determining infrastructure, but a precarious and never to be taken for granted balance of power. This, Gellner believed, was the key to the difference between Islam and the West. 'The difference would seem to be less in the absence of ideological elements than in the particular balance of power which existed between the various institutions in that society.'[14] We have 'In the polity, an unusual balance of power, internally and externally. . . .'[15]

This insight is synthesized and given coherent expression as the central theme of *Conditions of Liberty*. In the majority of agrarian societies, as in communism, nothing is separated, so 'political, economic, ritual and any other kinds of obligation are superimposed on each other in a single idiom.'[16] Feudal society in the West saw a partial separation. There was the start of a separation of religion and politics. Ancient society was 'eventually replaced by a new order, one in which the Christian separation of religion and polity made individual liberty thinkable.'[17] However, the process was a slow one, for Gellner believed that the political and economic were still fused together until feudalism collapsed.

> In feudal society, as political and economic strata are conspicuously visible and manifest, indeed are legally and ritually underwritten, it would seem everything is clear. There is no pretence. There is also no separation. There is only one social order, political and economic. There is no talk of Civil Society as distinct from the state.[18]

Yet, mysteriously, out of this unified world, emerged something new, a separated world. This is the world of 'Civil Society'.

The peculiarity of the separation, and the fact that its implementation hung in doubt in the latter half of the eighteenth century, he noted as follows. 'Civil Society is based on the separation of the polity from economic and social life . . . but this is combined with the absence of domination of social life by the power-wielders, an absence so strange and barely imaginable in the traditional agrarian world, and found so surprising and precarious by Adam Ferguson.'[19] The separation of politics and economics became entrenched and 'this separation is an inherent feature of Civil Society, and indeed one of its main glories'.[20] Indeed, this is the defining characteristic of Civil Society, which 'refers to a total society within which the non-political institutions are not dominated by the political ones, and do not stifle individuals either'.[21] The separation is complete. 'The emergence of Civil Society has in effect meant the breaking of the circle between faith, power and society.'[22]

In many ways this is a cogent restatement of many of the insights of earlier thinkers. What it special is not the thoughts in themselves, but the fact that they were written recently, during the second half of the twentieth century, when most of those around him had forgotten the riddle. Gellner restated the earlier vision in a new form. The new world has become so much part of the air we breathe that the shock felt by Montesquieu, Hume, Smith and Ferguson, or of comparative strangeness best exemplified in Weber, has been forgotten by most of us. Islam and the Soviet bloc, and perhaps memories of Czechoslovakia before the Second World War, constantly reminded Gellner that none of this is to be taken for granted, that it is indeed not the 'normal' condition of man.

A living experience of different worlds also made Gellner more aware than many of the cost of disenchantment. The insulation of various spheres of life has its own costs as well. Although it allows people to think freely and to act rationally it is, of course, caught in the deeper contradiction that the real world is not separated into watertight compartments. We have to believe that religion and politics, morality and economics, kinship and politics are separable and can live amicably alongside each other. But the garment is thereby torn apart arbitrarily; reality is a seamless web, as people living in the majority of human societies have realized. Marx recognized this in his concept of alienation, Durkheim in anomie, Weber in disenchantment. Gellner adds his own voice in elaborating these contradictions.

Based on his experience in Islamic and communist societies, and his reading of history and anthropology, Gellner suggested that if we looked at the last ten thousand years of human activity we could discern a powerful law which seemed to govern agrarian societies. The law was that they were bound to hit a ceiling where political violence curbed economic growth. Gellner put this law as follows. 'Material surplus generally, though not universally, makes for political centralization. And although political power and centralization in agrarian society is fragile, often unstable, it is nevertheless extremely pervasive.'[23] This is because

> The moment there is a surplus and storage, coercion becomes socially inevitable, having previously been optional. A surplus has to be defended. It also has to be divided. No principle of division is either self-justifying or self-enforcing: it has to be enforced by some means and by someone.[24]

It is also the case that 'Wealth can generally be acquired more easily and quickly through coercion and predation than through production.'[25] Consequently we find that in agrarian societies it is the warriors who are the highest group: 'specialists in violence are generally endowed with a rank higher than that of specialists in production'.[26] It is a world of competition, violence and scarcity. Thus 'Roughly, the general sociological law of agrarian society states that man must be subject to either kings or cousins, though quite often, of course, he is subject to both.'[27]

From our vantage-point at the end of the twentieth century, we can see that this is a tendency, rather than an iron law. There have been temporary and short-term exceptions, but Gellner's chief interest was in the major exception when something very unusual happened.

> Certain societies, whose internal organization and ethos shifted away from predation and credulity to production and a measure of intellectual liberty and genuine exploration of nature, became richer and, strangely enough, even more effective militarily than the societies based on and practising the old martial values. Nations of shopkeepers, such as the Dutch and the English, organized in relatively liberal polities, repeatedly beat nations within which martial and ostentatious display, dominated and set the tone.[28]

This is the miracle, and it happened in north-western Europe, at the very time when the Enlightenment thinkers started to analyse it. 'Once only did the balance change definitively, under exceedingly favourable circumstances – eighteenth century England...'[29]

Not only was there sustained economic growth, but the natural tendency towards growing absolutism and greater stratification and the suppression of free thought, were all simultaneously broken.

The escape from the domination of thought by the political and religious powers was extraordinary.

> The dependence of the individual on the social consensus which surrounds him, the ambiguity of facts and the circularity of interpretation are all enlisted in support of the fusion of faith and social order. This is the normal social condition of mankind: it is a viable liberal Civil Society, with its separation of fact and value, and its coldly instrumental un-sacramental vision of authority, which is exceptional and whose possibility calls for special explanation.[30]

Equally strange was the escape from the tendency towards 'caste'. Thus the 'astonishing egalitarianism of modern society ... has inverted the long-standing and seemingly irreversible trend of complex societies towards ever-increasing social differentiation and accentuated, formalized hierarchy'.[31]

Of course the escape may only be temporary, just as it is fragile. The open and expanding society was very nearly snuffed out by the Second World War. Only very recently has it become obvious that the other option, communism, is unlikely to take over the world.

Thus Gellner had specified a puzzle, the exit of one part of the world from the apparently closed circle of agrarian political systems. He saw the sociologist's central concern as the need to 'explain the circuitous and near-miraculous routes by which agrarian mankind has, *once only*, hit on this path; the way in which a vision not normally favoured, but on the contrary impeded by the prevailing ethos and organization of most human societies, has prevailed ... it is most untypical. It goes against the social grain.'[32]

*

One condition of the exit from agrarian civilization, lies in the development of religion. Like Weber, Gellner did not suggest that Protestantism intentionally or directly caused capitalism. Firstly the famous ascetic virtues of hard work, honesty and accumulation were an accidental by-product of the Reformation. Part of what Protestantism did was to push to one extreme a general tendency in much of western Christianity towards an attack on a magical and ritual embeddedness. Some of the explanation for the early growth of an unusual thought style in the West lies in Christianity, that is to say 'the impact of a rationalistic, centralizing, monotheistic and exclusive religion. It is important that it was hostile to manipulative magic and insisted on salvation through compliance with rules, rather than loyalty to a spiritual patronage network and payment of dues.'[3] Gellner outlined this great transition which occurred over 2000 years ago with the development of Christianity out of Judaism.

Over time, this asceticism, the tension between the material and spiritual world, tended to become overlain in Catholicism with a world of miracles and magic. Protestantism was the extreme attempt to restore it to its original anti-magical cleanliness.[34] This movement towards a 'disenchanted' world is an ideal background for orderly science and orderly capitalism.

Gellner argued that two revolutions were needed. The first was to separate thought from the material world and put it into the hands of the clerisy. The second separation, between the forces of coercion and those of cognition, between rulers and clergy, was equally important. He argued that 'It is hard to imagine perpetual and radical cognitive transformations occurring in a society in which the old alliance of coercive and clerical elements continues to prevail. They would suppress and smother it.'[35] How then did this second revolution occur?

Gellner suggested that something odd happened in the seventeenth and eighteenth centuries, which set thought free from its previous embedding in politics. For example, 'Descartes proposed and pioneered the emancipation of cognition from the social order: knowledge was to be governed by its own law, unbeholden to any culture, any political authority.'[36] This first emancipation, which the Counter-Reformation had tried to crush, only became firmly established after the eighteenth century Enlightenment. Religion restricted its claims to areas which 'do not prejudge the results of free and empirical inquiry'.[37]

Puzzling on how mankind escaped from the joint domination of priests and kings, Gellner developed the idea that it was because the clerics and the rulers fell out with each other. The 'normal' situation in agrarian civilizations was described by Durkheim, who 'sketched out what is really the generic social structure of agroliterate societies, namely government by warriors and clerics, by coercers and scribes'.[38] Yet this Caesaro-Papist concordat, the tension between Church and State is a peculiarly western characteristic as compared, for instance, to India or China. Gellner quotes Hume's explanation for the toleration in England and Holland; 'if, among

Christians, the English and the Dutch have embraced principles of toleration, this singularity has proceeded from the steady resolution of the civil magistrate, in opposition to the continued efforts of priests and bigots.'[39] But why were the civil magistrates opposed to religious extremism?

The key, Gellner suggested, may have been in the stalemate between a powerful Church and a powerful State, both seeking a monopoly yet neither able to obtain it.

> The separation of, and rivalry between, these two categories of dominators may well constitute one of the important clues to the question of how we managed to escape from the agrarian order. Priests helped us to restrain thugs, and then abolished themselves in an excess of zeal, by universalizing priesthood.[40]

It also appears likely that the religions of both Holland and England represented reactions to Rome, conceived of as a foreign, dominating institution.

The second thread of Gellner's explanation lies in the relation between the political and the economic.[41] His first premise, as we have seen, is that as societies develop into what we call 'civilizations', predation (politics) will dominate production (economy) and constantly restrict its development. It is a kind of Malthusian law of power. If through some accident or discovery, wealth is increased, it will lead to a rise in predation which will force mankind back to that world of violence from which momentarily it seemed to be freeing itself.

Of course, from time to time the relations of production and predation are reversed, and there is a period of economic and cognitive growth, as in Greece or the Italian city-states.

> Under favourable circumstances, power had very occasionally moved from thugs to traders even in earlier periods: but as long as there was a kind of ceiling on economic development, the shift did not proceed too far and either reached a limit beyond which it could not go or was eventually reversed.[42]

In general, looking over the long history of mankind up to the middle of the eighteenth century, it seemed true that 'political considerations trumped economic ones and the economic side of life simply could not be granted full autonomy – in other words, a market society was impossible – because the economy was so pathetically feeble'.[43] The normal tendency was for wealth-producing oases to be overrun by the surrounding military powers, as happened in Italy, southern Germany or the Hanseatic League. 'Commercial city states are a fragile rather than a hardy plant. Why should the free merchants of north-west Europe fare any better than their predecessors who lie buried in the historic past?'[44]

How was the 'stability or stagnation of productive forces – which, all in all, applies to agrarian society... eventually replaced by a permanently growing economy'?[45] Adam Ferguson had noticed, like Adam Smith, that it was happening, yet 'He does not adequately analyse the distinctive conditions which have led in modern north-west Europe to the subordination of coercers to producers.'[46] He does not explain how it was that 'under the new dispensation, the relative attractiveness of production and coercion changed. It is no longer more honourable to become rich by warfare rather than by trade'.[47] The subduing of political by economic power was the great triumph. 'Marxism made it a taunt that the bourgeois state was merely a kind of executive committee of the bourgeoisie: that this should ever have become possible is perhaps mankind's greatest social achievement ever.'[48]

Gellner explained the sudden dramatic switch by invoking a new, special factor, namely the development of technology and science. As the change began, the important thing was that there was technological growth, but that it was not too obvious.

> So early development may well have depended on the relative feebleness rather than the power of innovation. In fact, by the time the new world emerged in full strength, and its implications were properly understood, it was too late to stop it. It had been camouflaged by its gradualness, and that was made possible by the relatively non-disruptive nature of its techniques.[49]

Smith and Ferguson's pessimism had been well founded given the history of mankind. Yet both Smith's economic and Ferguson's political pessimism 'came to be invalidated by the same factor, by the tremendous expansion of productive power consequent on the impact of scientific technology'.[50] In the eighteenth century, a phenomenon whereby 'commerce and production... take over from predation and domination' for the first time in history perpetuated itself because it was 'accompanied by two other processes – the incipient Industrial Revolution, leading to an entirely new method of production, and the Scientific Revolution, due to ensure an unending supply of innovation and an apparently unending exponential increase in productive powers.'[51] Thus the 'entire shift from valuation of coercion to valuation of production was only possible because, rather surprisingly, indefinite, sustained, continuous technological and economic improvement *had* become possible'.[52]

A method was devised by which a country could rapidly become rich by increasing production, which meant that it was also able to became politically dominant. Technological expansion became a virtue, rather than a threat. The successful were not those who pursued the straight path of predation, but those who put much of their energies into production. 'Astonishingly, the regime in which oppression and dogmatism prevailed was not merely wicked, but actually weaker than societies which were freer and more tolerant! This was the essence of the Enlightenment.'[53] Thus '*sustained and unlimited* expansion and innovation . . . finally turned the terms of the balance of power away from coercers and in favour of producers. In the inter-polity conflict, no units managed to survive and to continue to compete if their internal organization was harsh on producers and inhibited their activities or impelled them to emigrate.'[54]

Thus the 'fittest' were now those who espoused that mix of openness and technological progress whose model was England. 'The economic and even military superiority of a growing society then eventually obliged the others to follow suit. Natural selection secured what rational foresight or restraint had failed to bring about.'[55] In pursuing this argument, we can see Gellner considering themes which were elaborated by Montesquieu and Adam Smith. The great difference is that Gellner can see the longer-term outcome, can add the Industrial Revolution, and can even see a modern rerun of the process in the collapse of communism in the face of the open capitalist West.

Why then did the change occur first in western Europe? Here Gellner elaborates a theme which also echoes the Enlightenment theorists. It could happen because Europe was split into a number of medium-sized states. Usually an improvement in technological power will strengthen domination

> But in Europe the process was taking place within a multi-state system, and the thugs were unable to use growth to strengthen

themselves everywhere at the same time and to the same extent. The various thug states were also engaged, as was their habit and joy, in conflict with each other. Those which had tolerated or were for one reason or another obliged to tolerate, prosperous and non-violent producers in their own midst, suddenly found themselves *more* powerful – because endowed with a bigger economic base – than their rivals.[56]

In huge absolutist empires, predation will eliminate production. 'But in a plural state system, in which other states prosper dramatically and visibly, the throttling and throttled systems are in the end eliminated by a social variant of natural selection. In a multi-state system, it was possible to throttle Civil Society in some places, but not in all of them.'[57] The continuous growth produced by science and technology not only provides an adequate 'bribery fund' to buy off the powerful, but it will also make it possible to solve the problem of keeping people in order without naked force. Thus it is also the basis for democracy. 'Only in conditions of overall growth, when social life is a plus-sum, not a zero-sum game, can a majority have an interest in conforming even without intimidation.'[58]

*

Yet Gellner's solution only picked up about half of the Enlightenment argument. It embraced the philosophical side, but only parts of the historical. One might think that this was because he was not an historian and this, of course, is part of the reason. But there is more to the omission than that. What exactly was missing can be seen if we compare his solution to that put forward by earlier thinkers. Gellner specified the problem well, and saw that the solution lay in a theory of structural balances. But all the middle part of the Enlightenment solution, the contrast between China and Europe, the analysis of the Roman failure, the nature of the feudal contract and the feudal gate, the loss of the balance in much of continental Europe, the peculiar case of England and why it developed differently, and the consequences in America, all this is missing. He only picked up the Enlightenment clues again when he discussed the way in which the rich become the powerful, and in relation to the costs and dangers of the move from agrarian society.

There are several reasons why all of this middle section had become invisible even to an observer as acute as Gellner. Firstly, while he rejected much else, he did accept the basic evolutionary model that had developed in all of the social sciences. Its attraction for him was all the greater because it is, in fact, a simplified version of one aspect of the Enlightenment synthesis. The 'classic' four-stage theory of some of the Enlightenment thinkers, the hunter-gatherer, pastoral, settled agricultural, commercial, was simplified by Gellner into three stages, tribal, agraria, industria. This also roughly fitted with the growing anthropological divisions into 'tribesmen', 'peasants' and 'modern'. Although Gellner was well aware of the difficulty of the move from agraria to industria, he accepted the 'before', 'during' and 'after' model. All societies were 'agrarian' up to the seventeenth century; then they began to be transformed. And basically all 'agrarian' societies were structurally similar – they had a 'normal' shape which he described in detail, a sharp hierarchy, position based on status, domination by lords and priests and so on.

Quite early in his intellectual life Gellner seems to have convinced himself that three-stage models are the best and that world history can be fitted into such a model. This became a dogma in his book on nationalism: 'Mankind has passed through three fundamental stages in its history; the pre-agrarian, the agrarian and the industrial.' Or again he writes, 'My own conception of world history is clear and simple: the three great stages of man, the hunting-gathering, the agrarian and the industrial, determine our problems but not our solution.'[59] Trinitarians who subscribe to the 'elegant and canonical three' stages (Comte, Frazer or Karl Polanyi) are praised. In an interview in 1990 he admitted that 'What is true is that I very much like neat, crisp, models, and try to pursue them, and I would be very uncomfortable if I didn't have one.'[60]

The difficulty is that such a model, if taken as a universal law of development, does determine not only the problems, but also the nature of the solutions. If we believe with Gellner that there are these three types, each distinct and different, it is indeed difficult to see how the movement from one to the next occurred. Attractive as three-stage theories are, they are probably an 'idol of the mind' in Bacon's sense. They are useful as organizing devices, showing some strong tendencies. But they are not laws of progress. We should treat all ideas of stages as, at the most, tendencies, as gauges against which we measure actual histories. If reified into necessary sequences and laws of development, they blind us to what actually happened.

Particularly significant for us here is that part of Gellner's scheme dealing with the 'middle' stage of 'Agraria'. There is here a tremendous lumping together of differences in 'Agraria'. Here we seem to have everything from pastoral nomads to densely settled India and China, almost every conceivable kind of kinship system, numerous variations in religious and political organization. If they are all lumped together or generically similar, it makes the emergence of modern industrial civilization inexplicable. I suggested that once we allow for the possibility that, say, fourteenth-century England, though 'agrarian', was very different from fourteenth-century Bohemia, Ghana, Peru or China (or the approximate places where these names would later apply) then it becomes easier to assess what may have happened. Gellner in his reply to this suggestion reasserted the structural similarity of all agrarian societies, a similarity which means that 'Agraria is doomed, by the very logic of its situation, to remain what it is'.[61]

One consequence of homogenizing the agrarian 'stage' can be seen if we examine Gellner's treatment of feudalism, the main area where he failed to follow the Enlightenment trail. Gellner realized that one of the quintessential features of modernity lies in its peculiar blend of status and contract. Gellner realized that modern civilization is based on the coexistence of both principles. A modern Civil Society has to have at least, temporary, flexible, communities, as well as individual choices. 'Civil Society is a cluster of institutions and associations strong enough to prevent tyranny, but which are, none the less, entered and left freely, rather than imposed by birth or sustained by awesome ritual.'[62] In a central passage he pointed out the tensions, peculiarities and contradictions. Modern man

> is capable of combining into effective associations and institutions, *without* these being total, many-stranded, underwritten by ritual and made stable through being linked to a whole inside set of relationships, all of these being tied in with each other and so immobilized. He can combine into specific-purpose, *ad-hoc* limited association, without binding himself by some blood ritual.[63]

This is the peculiarity, the existence of a combination of all those nineteenth-century dichotomies – Community and Association (Tonnies), Status and Contract (Maine), Mechanical and Organic solidarity (Durkheim) and so on.

There is in fact a partial, but only a partial movement along these dichotomies.

It is *this* which makes Civil Society; the forging of links which are effective even though they are flexible, specific, instrumental. It does indeed depend on a move from Status to Contract: it means that men honour contracts even when they are not linked to ritualized status and group membership. Society is still a structure, it is not atomized, helpless and supine, and yet the structure is readily adjustable and responds to rational criteria of improvement.[64]

By some miracle, 'these highly specific, unsanctified, instrumental, revocable links or bonds are effective! The associations of modular man can be effective without being rigid!'[65] The small company or football team or orchestra are examples of this. Yet the ability to hold people together and yet also give them freedom is very unusual.

Gellner did not realize that there was something odd about feudalism, and particularly the form that developed in England. While recognizing that the 'relationship between members of various levels in this stratified structure ... are ... ideally and in principle, contractual' and 'even affirms a curious free market in loyalty', Gellner still believed that feudalism is 'governed by status and not contract'.[66] Thus he can compare a modern 'open, mobile, growth-oriented, modular social order' to a 'feudal or baroque' one, which is 'absolutist, status-oriented, anti-productive'.[67] It is thus difficult for him to see how strange and powerful feudalism was. If the major transformation which Gellner analyses is rephrased in other terms as the movement from status-based to contract-based societies, or from *Gemeinschaft* to *Gesellschaft*, then, according to Gellner, feudal societies are still 'governed by status' and hence on the wrong side of the 'great divide'.

Yet the greatest thinkers on this subject are united in placing feudalism on the 'modern' side of the great divide. Montesquieu, Adam Smith and Tocqueville were all aware of the deeply contractual nature of feudalism. Their intellectual descendants, for example Sir Henry Maine and F.W. Maitland, re-emphasized this surprising fact. It is worth repeating Maitland's famous comment on Maine: 'The master who taught us that "the movement of the progressive societies has hitherto been a movement from Status to Contract" was quick to add that feudal society was governed by the law of contract.' Maitland added his endorsement: 'There is no paradox here.'[68] In other words that very element of 'progress' and 'growth' which Gellner singled out is present in feudalism. Not only, as Gellner realized, was religion separated from politics, but politics and economics were already in a contractual relationship to each other. We already have the peculiarity he is searching for well before the eighteenth century.

Once we have accepted that the essence of feudalism is its contractual nature, and that this flexibility was widespread in the period after the fall of Rome, the puzzle becomes, as Montesquieu and Tocqueville realized, how to explain the fact that gradually over most of Europe, with the notable exception of England, contract turned back into status. Much of their work helps to solve Gellner's puzzle by showing that for peculiar reasons a contractual, relatively open world was preserved in England amid an advancing sea of 'caste' and political absolutism.

Gellner's third assumption lies in relation to his prime mover. For Gellner, as we have seen, the external factor which changed the world was the growth of science and technology. He seems to accept that they would grow naturally, as long as the conditions were appropriate. What he has done is to substitute 'science and technology' for Smith's driving mechanism, namely the division of labour. Indeed there is hardly any substitution, for Smith himself envisaged the growth of technology and knowledge as important constituents of the increasing division of labour.

If it could be assumed that science and technology will naturally grow if the brakes are taken off, Gellner's solution would be plausible. To use one of his favourite metaphors, all that was needed was to 'unthrottle' the system and release the negative forces which prevent 'natural' growth. If one allowed production a free rein, then the rest follows from the 'natural course of things'. Such an assumption means that Gellner's attention was focused on the traps and negative factors.

Yet we know that the puzzle is deeper than this. Possibly peace, easy taxes and justice, which can be read as shorthand for that separation of politics, religion and economics which is at the heart of modernity, are indeed necessary factors for sustained technological and scientific development. But we know that they are not sufficient. There are many counter-examples through history where, for instance, as in Tokugawa Japan, there were long periods of peace, relatively easy taxes and a firm and universal judicial system. Yet technology and science remained almost stationary. Something more is needed.

To our benefit, and with characteristic wit and width of vision, Ernest Gellner enumerated some of the 'conditions for the exit'. But by letting his mind rest, by invoking a 'natural tendency' for growth, all else being equal, he was unable to solve the riddle of modernity.

Gellner's solution to the puzzle, which he saw so clearly, lay halfway between the Englightenment and the Marxist answer. It had the surprise and contingency element of the Enlightenment, but added the total transformation of modes of production of the Marxist approach. Basically, what happened was an amazing, surprising, unlikely breakaway of parts. Given his foreshortening of history and lumping, there was less chance of seeing earlier roots and continuities. For if everything was basically one lump, the chances of finding a solution to the emergence of something different were slim. If all was the same, why a sudden shift?

Gellner admits that when dealing with such an improbable, contingent and complex set of events it is very unlikely that one will find an entirely satisfying 'solution'. 'The origins of industrial society continue[s] to be object of scholarly dispute. It seems to me very probable that this will continue to be so for ever.'[69] The 'first miracle had occurred when men for obscure reasons persisted in working a set of levers not yet known to work'. So that 'on one occasion and in one area, the message did prevail, thanks to very special circumstances: and the world was transformed for good'.[70] These remarks successfully capture the essential point about the uniqueness and lack of inevitability of the process. Miracles are as difficult to explain as accidents. 'The notion of a unified orderly Nature and an egalitarian generic Reason led, by a miracle we cannot fully explain, to an effective exploration and utilization of nature'; yet Gellner does attempt to explain the inexplicable, while implicitly recognizing the impossibility: 'We have striven to explain how one society, and one only had, by a series of near-miraculous accidents...'[71] escaped into modernity.

In his last months he continued to show his puzzlement. It had been pointed out that his approach 'makes the emergence of modern industrial civilisation inexplicable', to which he replied 'It is', drawing attention to Adam Smith's bafflement.[72] He continued by arguing that while 'Agraria is doomed, by the very logic of its situation, to remain what it is. We know, in fact, that we have broken out of it: if the argument showing that this cannot be, has some cogency – which to my mind it has – then we must be puzzled concerning the nature of the explanation.'[73] He knew that it could not happen by accident, but nor can it happen by design. He agreed that the balance of powers is at the heart of the matter, but

> continue[s] to think that the conditions of Agraria militate against it so that an explanation over and above the random play of factors is required if it does happen ... The providential balancing out of powers or institutions ... is a luxury which agrarian society cannot allow itself. It is not allowed to happen even by accident.[74]

Thus, right to the end, he was faced with an event which could not, should not, yet did happen.

14
The Riddle Resolved?

In the previous chapters I have looked at some of the most notable attempts to solve some of the problems which I have termed the 'riddle of our world', that is the growth in the capacity of humans to produce resources, culminating, after 10 000 years, in a sustainable escape from agrarian civilization. All these encounters have thrown light on different aspects of the puzzle, seldom contradicting each other or differing on essentials. Indeed they show a surprising agreement and continuity of approach, while each adds a set of new insights to supplement the others. In this chapter I shall try to synthesize their insights into one composite answer to the riddle of the modern world.

Our thinkers are united in their specification of what the central problem is. They agree that human beings are creative, inventive, curious, often motivated by strong drives to better their position. In appropriate conditions they will tend to increase their manipulation of the natural world so that their standard of living rapidly improves. They have the potential for cumulative or non-linear growth in their ability to produce resources. Indeed, for short periods in their histories, many regions or civilizations have seen such a growth.

On the other hand, experience showed them that the majority of such periods of growth came to an end quickly and that long periods of stasis or even decline were the norm. Growth was exceptional, stasis was the usual condition. Thus there must be a set of very powerful, negative, forces which crush man's natural abilities and desires. Their concern was to specify these constraints or traps and to show how they had operated and sometimes been avoided for limited periods.

They were well aware that as the potential for rapid growth became greater, through higher levels of knowledge and technology, so likewise the negative pressures grew at an equal or greater pace. As each form of civilization succeeded the previous one it faced new and more significant problems. This can be seen if we look at the extremes. To move from hunter-gatherer to tribal societies required a relatively minor shift – domestication of plants and animals. The checks on this were relatively light, though starting at a subsistence level with practically no technological support, the transition was immensely difficult. The push was weak, and the counter-push was also quite weak. The two were well enough balanced to prevent any change in most of the world for over five hundred generations of human existence. In Australia, over 300 generations of hunter-gathering never led to anything different before the arrival of white colonists.

At the other extreme, if one took a great empire like China, it was possible to see how both the potentials for transformation and the negative pressures were huge, and again just about balanced each other. The technological, intellectual, cultural and social sophistication of China by the fourteenth century was immense, far ahead of Europe. It had developed a knowledge of almost all the techniques necessary for industrialization, it had a very sophisticated and literate ruling group; it was peaceful and orderly. People were generally hard-working and profit oriented. Yet 400 years later, apart from the undoubted success in feeding a much larger population, it had made no noticeable technological, scientific or social 'progress' and was now 'falling behind' Europe.

Nearer at home, the greatness of the Roman empire, heir to all of Greek science and with its own developed organizational technologies, had collapsed, and more recently the promise of the Hapsburg Empire, of Renaissance Italy and southern Germany or even *ancien régime* France, had faded away or reached a plateau.

The potential of all these civilizations for rapid cumulative transformation was immense. Millions of hard-working, ambitious, inventive citizens surrounded by a wealth of practical, reliable, knowledge of how to manipulate the natural world to their own uses should have gained in opulence from generation to generation. The fact that they did not do so shows the strength of the negative pressures.

Much of the thought of our informants is concerned with these negative pressures and how, occasionally, they were overcome. Their central understanding was that as productive technologies grew in power, they were more than counterbalanced by predatory tendencies, which began to halt productive growth. Within these predatory tendencies they included not only obvious external predation, warfare and raiding of others, which often checked a civilization, but equally important, internal predation, that is to say the predation of priests, lords, kings, and even over-powerful merchant guilds. This internal predation usually took the form of increasingly sharp stratification – castes and estates – and increasingly absolutist religion and government and hence the destruction of personal liberty of action and thought.

The process within agrarian societies was a circular one. As productive technology produced greater surpluses, these almost automatically increased predation by increasing temptations. Success created envy and smaller states or cities were destroyed by neighbours. Even huge civilizations such as China or India or eastern Europe were devastated by predating Mongols. A perpetual levelling took place. Likewise the growing wealth led to the temptation to expand and conquer and the centre was finally ruined by the burden of imperial dreams, as had happened in Rome, the Hapsburgs or Louis XIV's France. Almost automatically surpluses generated aggressive behaviour. And such militaristic activities directly led to the twin forms of internal predation – higher taxes, rents and social stratification, and increasingly absolutist power and political predation.

This was the central trap, supplemented powerfully by the Malthusian tendency for rapid increases in population to outstrip all growth in production and hence to add famine and disease to war and internal predation as checks on sustained growth. This was the agrarian trap, and every great civilization up to the seventeenth century had finally become entangled in it and either collapsed or, like China and Japan, become immobile.

The riddle to which our thinkers addressed themselves was how one escaped this apparently inevitable fate. During their lifetimes they speculated on the growing realization that against all the apparent predictions and laws, an escape to something else was indeed happening.

All our thinkers looked at the process from the outside. From France or Scotland they increasingly focused on a new world which seemed to be emerging first in England and then in America. The natural tendency towards cumulative growth inherent in man's intelligence and nature had always previously been checked by the iron laws of population and predation, which had finally crushed productive increases. This was the first contradiction. Something was happening, and towards the time that Tocqueville wrote, had clearly happened, in England and America, which showed that the iron laws were not laws at all, but just powerful tendencies which could, apparently, be avoided. The difficulty of avoiding them was immense, as the history of all previous civilizations showed, yet a set of peculiar balances might be achieved for long enough to do so. How these mechanisms occurred and worked was the riddle which they sought to answer.

There seems to be a consensus among all our informants that an answer to the riddle must lie in the balance of forces. They were all aware that a structural solution, that is to say one which focused on the *relations* between the parts, rather than the parts themselves, was necessary. The key to the mystery lay in the difference between the normal tendency, which was towards a certain set of interlocked and rigid institutions, and the exceptions, where the parts remained independent, antagonistic even, and hence flexible.

Putting this more explicitly, they suggested that the normal tendency was as follows. In tribal societies, almost everything was encompassed within kinship – political power, religion, economy were all embedded within this. Hence economic or political development was severely constrained. To change one element was to attack them all. The development of civilization depended, to a certain extent, on the weakening of kinship (status) and the growth of the power of other institutions – the economy and technology, the political structure (state systems) and religion (universalistic religions of the book). This was the huge leap and it allowed a freeing of energies and growth in all forms of production.

Yet there seemed a powerful tendency in agrarian civilizations for the structure to solidify again, this time into a new form of overlapping and dominating structure usually based on an alliance of priests and rulers. As productive wealth increased, the tools of power, both military and ideological, increased proportionately. The history of almost all civilizations, or periods within ancient civilizations such as China, was for a period of anarchy and confusion, where productivity was low but flexibility high, to settle down into higher productivity but declining flexibility. A clear example of this lay, they thought, in the history of western Europe, as most of the continent moved from a lightly populated, technologically backward, but mobile, flexible and highly contractual feudal world which covered the continent from the sixth to twelfth centuries, into increasingly rigidified, status-based, politically and religiously absolutist civilizations from the thirteenth to eighteenth centuries.

They argued that as part of the swing from production to predation, there was increasing domination of all life by an increasingly closed world of politico-religious power which had crushed kinship, or suborned it to its use (as in China) and tried to maximize short-term, and even immediate benefit (as in war) from the productive labours in the economy. As the gaps, tensions and balance between institutions were closed, the space for technological and productive growth was increasingly reduced. Indeed, any advance in the wealth of producers – whether craftsmen and manufacturers, or merchants, or even peasants, was a distinct threat, as well as an opportunity for predation, and hence quickly crushed. Likewise any growth in intellectual production outside the central circle of power was potentially undermining and quickly put down as heresy.

In a sense we can see the development as a tendency towards centralization and inequality, of concentric rings of power and status, as the Sun King or God-Emperor, in which all countervailing forces or relatively independent centres of production of artifacts or ideas were crushed. Uniformity, homogeneity, a rigid and level landscape emanating from the centre, where all forms of activity were again made interdependent, as in the Confucian parallel between religious, political and kinship loyalties, this was the growing tendency. The weight of the fruit of increased production increasingly brought down the tree. Or, to use a mechanical analogy, a negative feedback loop was in operation.

That this was a natural tendency was not surprising. All were agreed that alongside sexual and intellectual drives, the desire to dominate and exert power over others was a basic human instinct. Indeed, much of human progress had arisen from the energy which this desire prompted. But the desire was ultimately selfish. Each individual would try to maximize his or her own power, and perhaps that of his small co-ordered group, whether family or caste. With this powerful aim, and with increased wealth and technologies of domination, predation founded on an alliance of the rulers and the thinkers, Kings and priests, subjecting the rest (the workers and 'producers') to increasing pressure, was an obvious strategy.

Indeed it was a strategy which could even, plausibly, be argued to be in the general interest. In a world where three quarters of the Eur-Asian continent was subject to periodic devastation from the wandering tribes of central Asia, or more locally from neighbouring powerful states, it made sense to put a great deal of productive wealth into predation and counter-predation. The philosophy of Machiavelli epitomizes this world where offence was the best form of defence, where those who aspired to virtue, peace, equality and liberty, were soon devastated. Even Christianity, founded on a gospel of turning the other cheek, witnessed the Crusades, the Inquisition and the final defeat of the Islamic threat at Lepanto.

Yet desperate though the ravages of war could be, there were recognizably equal dangers in too much peace. This again was best shown by the history of China and Japan. Long centuries of peace, in both civilizations, when military expenditure was relatively small and there was fairly light taxation, and even a powerful control of disorder, led only to the stagnation of technology and economy. Of course this could partly be explained by the Malthusian tendency towards rapidly increased population. Or it could be explained by the encouragement of stratification and labour rather than capital and technologically intensive agriculture partly caused by the peculiarities of rice cultivation. In a sense these countries were a warning of the dangers of too much success. The climate and agriculture produced huge surpluses, there was little to struggle against, and the system rigidified. It was a phenomenon which some of our informants also noticed in contrasting the early fertility and abundance of Mediterranean Europe, with the need to struggle and produce in the Protestant north.

*

All 'advances' are costly to someone – for example the labour-saving technologies which were the foundation of the industrial revolution were made at the short-term expense of millions of workers. The move to the new sources of agricultural (horses, windmills, watermills) or military (longbows, guns), or ideological (printing) power were all equally a threat to vested interests. Most civilizations were inhibited by such interests, or partially incorporated them as a new means of control. They only tend to be accepted because to fail to do so would mean that the competition from elsewhere would crush one – a sort of intellectual, technological and cultural arms race.

As several of our informants pointed out, this appeared to be the great difference between Europe and China or Japan. The plurality of small states in Europe, autonomous but linked by a common history, religion and elite language, almost incessantly at war and when not at war, in fierce cultural and social competition was the ideal context for rapid productive and ideological evolution. There was enough in common for ideas and inventions to travel swiftly, there was enough variation for separate centres of innovation to cross-fertilize. Europe was, to use Gerry Martin's phrase, a large system comprising a network of 'bounded but leaky', autonomous yet competing, states, and it had been so for about a thousand years before the industrial revolution. The tendency to form a vast homogeneous empire, the dream of Charles V, Louis XIV, Napoleon or Hitler, was never realized. The political and actual geography and the level of communications, military and ideological technology made it impossible. Huge diversities of religion, kinship systems, culture, farming practices and craft production, continued to exist and was encouraged by large geographical and climatic differences over a small area.

China, of course, also varied considerably geographically, and even culturally. But it spread politically over a vast area and came to hold a mass of individuals within one system of thought and organization. The geographical differences were not supported by the religious and political differences, which would have encouraged and protected competition. At first this made technological development very rapid. The economies of scale and massive demand, set China on a course which put her far ahead of Europe. But it seems as if by the fourteenth century the variability had been largely used up. Thereafter a conscious decision to shut out the undermining influences from abroad by ceasing sea or land exploration was completely the opposite of the outward expansion created by the competition of European states.

In western Europe it became obvious that external predation through voyages of discovery and conquest, incorporating new ideas and technologies and peoples, was the way to wealth, as the Venetians, Portuguese, Spaniards, Dutch and finally the British discovered. So the internal variation and competition that had stimulated the first burst of productive creativity and had allowed the explorations to be effective, was supplemented by the enormously varied information from the civilizations of America, Africa, India and Eastern Asia. European states absorbed the wealth of their conquests, wrestled with the new knowledge and adapted and evolved their hybrid solutions very quickly. China and Japan closed their frontiers, for five centuries in the case of China, three centuries in the case of Japan, and suddenly discovered that their once superior, but now antiquated technology, was no match for that of America and the European powers.

Thus it appeared clear to our four thinkers that if we are to find a solution to the riddle then we must look to the relationships between political, ideological, social and economic power or, as they might have put it, the relations between liberty (political and religious), equality, and wealth. All of them saw the key to the mystery in a peculiar association between these manifest in England, which ran against the current that had increasingly led to a growth of rigid uniformity in most civilizations.

Furthermore they knew that the solution must face and answer two further problems. The first was how to avoid the previous law that increasing wealth inevitably brought nemesis, either through internal weakening and hence predation by outsiders, or through the urge to conquer others. Put in another way, how was it possible to achieve that mysterious balance which clearly the Dutch and English had achieved by the seventeenth century where it was possible to be both virtuous (liberal, relatively egalitarian, non-absolutist), highly productive (using almost all one's energy in devising improvements in knowledge and technology) and at the same time be militarily powerful. That virtue was not just its own reward, but brought other rewards, was the amazing new fact. Look after the Kingdom of God, aspire to resource expansion, create a balanced polity and a not too uniform legal or social system, and all else would follow. This was decidedly not the experience, except for short periods, of previous civilizations. How could Holland and particularly England and America do something which had eluded the Hanseatic league, or northern Italian city-states?

The second puzzle was how was it possible to overcome the contradiction between the nature of man which was based on 'private vice', and the obvious fact that increasingly complex societies have to be based on a vast amount of trust, co-operation, altruism, generosity. How, in Pope's phrase, could 'self-love and social be the same' become fused into Mandeville's 'private vice, public benefit?' All previous civilizations had seen the contradiction as leading, finally, to destructive and aggressive confrontations, or, where, as in Confucianism, self-love was banned, to apparent stagnation. European society tried to harbour and even encourage self-assertion, individual self-love, yet to temper it so that it gave strength to the whole, rather than shattering it. The English case was the area where all the theories could be explored at their fullest.

A study of Roman civilization showed clearly that it had abandoned its democratic ideals and moved into the usual hierarchical and absolutist trap. The roots of modern liberty and equality must lie elsewhere. Montesquieu famously found them in the 'German woods' where the wandering tribes who overran the Roman Empire returned western Europe to its democratic origins. They were warlike, individualistic, egalitarian. With a surprisingly developed money-consciousness and judicial system, they set up small, balanced kingdoms throughout the old Roman lands.

By the twelfth century western Europe was uniformly 'feudal', that is to say the basis of society was an 'artificial' feudal contract, an act of will, an agreement to concede certain economic and political rights. There were no strong birth-based differences. The centre was not too strong, kinship was secondary to politics and religion was significant but not too powerful. In other words, European states had achieved that unlikely balance between institutional forces, between centre and periphery, between the ambitions of rulers, priests and people, so that none became supremely dominant.

It was during this period of balance that there was rapid technological, artistic and intellectual growth. This was the era of the extension of horse and plough agriculture, the spread of water and wind power, the reception of Greek, Indian and Chinese science by way of Islam, the founding of universities, and the growth of towns and trade. Many of these developments occurred because of that very balance of forces. For example, Adam Smith noted the miraculous growth of towns and economy. Normally they would have been stifled, but by using their leverage in the struggle between the king and feudal lords, they exacted precious freedoms.

Although structurally very similar to the rest, England was the extreme example of a propitious balance. The inherent danger of early feudalism was that, as Bloch put it, it led to the 'dissolution of the state' and the constant squabbles of over-mighty barons. In England, largely because it was an island, the Crown became unusually powerful and a centralized, but not too centralized, monarchy emerged under the later Anglo-Saxon kings. The genius of the Normans and Angevins consolidated and homogenized this system so that by the thirteenth century England was a wealthy, powerful, well-governed land, with a rapidly growing technology, trade links, strong armies and booming towns. It was basically, like the rest of Europe, a

Germanic kingdom with an engrafting of a more developed centralized feudalism.

This phase of 'feudal' civilization, the five or so centuries between the eighth and thirteenth centuries, provided the possible gateway to a kind of civilization never before envisaged in the world. A large, diverse, agrarian civilization was broken into small, competing quasi-states, unified under Christianity, yet preserving their differences. The system was based on contractual ties and not birth differences, on achievement rather than ascription. There was an antagonism yet balance between the growing power of the Church and the State, but neither was dominant and they had not yet formed a close alliance. Most of whatever power kinship had once held had been destroyed. Commerce and towns were encouraged, as was technological innovation in the saving of labour in the difficult and population-scarce environment. Thought and recovered knowledge were relatively free of the jealousies of a powerful clergy.

However, the next 500 years saw the balance destroyed throughout continental Europe. The normal tendencies and traps of agrarian structures, the temptations and tools which growing knowledge and wealth provide, worked their usual consequences through time. The threat of war, the threat of 'heresy' (the Crusades, the Albigensian heresy), the desire to predate on the wealth-creators, all these were forces which tipped the balance.

The institutional mechanisms for the emergence of the absolutist state and church are well known. The Catholic Inquisition, the reception of an authoritarian and centralizing Roman law, the emergence of caste-like birth privileges for the 'nobility', the destruction of the 'liberties' of all intermediary bodies such as trade guilds and town governments, the rising size of standing armies and the central bureaucracy, all these were catalogued by our informants. In essence, European civilization moved away from a 'feudal' one based on the flexibility of 'contract', to an *ancien régime* one based on 'status'. The tensions and separations of spheres were lost. A centripetal force seemed to be at work – everything gravitated upwards and towards the centre.

However there were exceptions, both at the national or local level – the rise of free cities in northern Italy or Germany being good examples. But they were soon crushed and the tendency continued. By the eighteenth century, the usual barriers – war, famine, disease, an increasingly impoverished peasantry, a parasitic nobility, a conservative clergy, an arbitrary and despotic law, a large and enervating bureaucracy and the heavy taxes to sustain it, increasing predation on merchants and producers, all were widespread.

From a position where, five centuries earlier there had been the potential for immense and continuing growth, that very growth in its early stages had provided the usual negative feedback to bring about a high-level equilibrium. This was the situation which recent historians have christened the 'general crisis' of the seventeenth century. Although perhaps reaching a level of wealth and knowledge (thanks to Greece) slightly above that of any hitherto existing civilization, including China, western Europe was still far from the organizational and technological level which would enable it to cross the barrier into the cumulative and self-sustaining growth necessary to improve the lives of an ever increasing population.

Indeed, the position was worse than this because the new arrangement of institutional forces meant that any increase in wealth merely fed the negative obstacles to further growth by way of a Malthusian rise in population or a rise in the various forms of predation. The balance had swung to a situation where predation dominated and squeezed production.

By the late seventeenth century there were only two apparent exceptions to this. One was the richest (per capita) country in the world, namely Holland, the other was England. Holland exemplified the advantages of a balance. It had discovered that a liberal course – separating and balancing, encouraging political and religious liberty, decentralizing power, avoiding extreme stratification, all encouraged a rapid growth of wealth so that a tiny country with such a virtuous structure could defeat the greatest empire in the West. Yet there were reasons why it seemed unlikely that Holland would provide the lever to tip the agrarian world into something fundamentally new.

To start with, it suffered from the defects of size. It was really a city-state, with Amsterdam and a constellation of smaller towns and villages. It was not self-sufficient in food, it was deficient in resources of good land and particularly the essential fossil fuel, coal. Furthermore it was also starting to obey the universal law of predation – though in an unusual, twentieth century manner. That is to say, an increasing amount of its energy was being taken out of production and being invested in speculative trading and banking. It lived off its fleets and empire to the detriment of its wealth generation through internal production and innovation. Thus its keenest observer, Adam Smith, felt it had reached the limits of its economic potential by the late seventeenth century. Furthermore it was being squeezed on two sides by the French and English and thus subject to another trap of predation – the need to spend an increasing amount on defence.

*

England was the exception that somehow solved the riddle. In 1200 England was a centralized and reasonably well-governed example of the western European pattern. From this we might have expected that, combined with its rapid technological and productive development, it would have been moved by the normal tendencies evident over Europe towards a precocious form of the trap – heavily absolutist, stratified, and caught in the grip of an intolerant inquisitorial religion. Yet by the time that Montesquieu visited England in 1739 he believed it hardly resembled the rest of Europe.

What had happened in those 500 years to make the trajectories so different, and why had it happened? In essence the easiest way to explain the difference is negatively. The three great tendencies which had swept most of Europe had not occurred. Political absolutism and centralization, with a ruler above the law and no countervailing forces, had not become established. There had been times when the tendency asserted itself, famously under King John, Henry VIII, Charles I and James II. But, in each attempt at monarchical aggrandisement, it had failed and a short period of absolutist rule had led to a reaction; Magna Carta, the rise of the Elizabethan parliament and sale of Crown lands, the beheading of the king and the Whig restoration. As most European states saw the vestiges of democracy swept away, the power of the Commons grew until by Montesquieu's time the contrast was overwhelming.

The trappings and mechanisms of absolutism which had arisen over much of the Continent failed to develop. There was no large central bureaucracy, but rather the devolution of power through a complex of often voluntary and honorary power holders such as constables and the justices of the peace. There was relatively easy taxation, an arrangement jealously guarded by the Commons. There was no standing army and few hired mercenaries. The legal system bequeathed by the Germanic invaders, the English customary law, was retained and no formal reception of inquisitorial and despotic late Roman law occurred. All these are both signs and associated features of the absence of the normal tendency for power to corrupt.

Amazingly the country grew steadily wealthier, yet the wealth was spread, property was secure, people had rights, and the King remained below the law. It was a miracle, often at risk, but it happened.

Secondly, the tendency for the ecclesiastical power to increase and to enter into a pact with the lay authorities never occurred. There remained a deep conflict between state and church. Neither was strong enough to subdue the other. From Becket to the destruction of the monasteries, the Crown kept the church in check. Both conspired to ward off the sweeping resurgence of clerical power in the fourteenth-century rise of the Inquisition or the sixteenth century Counter-Reformation on the continent of Europe. Instead of coalescing into a fervent and conservative force, England saw the rise of pre-Reformation independent sects such as the Lollards, then the Reformation itself which placed the individual at the heart of his religion and reduced the mediation of the theocracy, and finally the growth of sectarianism and a balanced toleration whereby religion became a private matter, seen in its extreme form in America. The final separation of religion and politics, rather than their coalescing, had been achieved, and religion also withdrew from economic life – or left such a life to the private conscience. Again, a tendency so manifest in Islam, Confucianism and Catholicism, to embed all of life within 'religion', had not occurred.

The third great absence was the tendency towards internal social predation, taking the normal form of increasingly rigid, caste-like, barriers between birth-given orders. As Tocqueville in particular brilliantly showed, the normal division into a high nobility, inferior bourgeoisie and crushed peasantry, did not occur. A hitherto unprecedented social structure emerged which was based on wealth and achieved status rather than blood. It was still very inegalitarian *de facto*, but the possibility of movement within it was great. Furthermore, the bulk of the population were placed at the middle level rather than there being a few immensely privileged, and a great mass of illiterate agrarian producers. The peculiar rank of the English gentleman, the exceptional status of 'yeoman', the high position of merchants and craftsmen, the prosperity of country dwellers, all these were signs of something peculiar, a proto-class rather than proto-caste or estate system. Such a wide spread of wealth, power and liberty obviously both reflected and affected the possibilities of political absolutism. It was only matched, in a slightly different form, in Holland.

The question still remains, though, of why, starting from a fairly similar origin in Anglo-Saxon civilization, England had gone against the normal tendencies. Here our informants developed two interlinked theories. One, developed by Montesquieu and expanded by Smith, concerned the hidden effects of material wealth and in particular trade and consumerism. In its crudest form, the theory was that, all else being equal, if a country could move into a long period of material growth, the tendency of the lords and the church to gain power would be deflected by greed. The lords would prefer consumer goods rather than retainers, and hence lose their military bargaining power. Likewise the Church through greed would lose the love of the people as it assembled treasures upon earth and end up like the monastic orders, stripped of almost all influence. The difficulty, of course, was to sustain growth at such a level and for long enough that this conversion of the powerful to the preferability of modified predation, or even dabbling in production, could be achieved. Often it happened for short periods and in limited areas, as in the environs of a city such as Florence or Sienna. But such a haven was soon exposed to marauders from outside and would revert to the usual predatory dominance of lords and priests.

Thus this theory was linked to the one unique feature of England, that it was a large enough island, just far enough away from a sophisticated continent. It was large enough to defend itself and to generate a great deal of internal diversity and trade. It was far enough away to make it difficult to attack or even threaten as long as it was defended by a good fleet. Yet it was not too far away, twenty miles rather than the 100 miles which made Japan so much more isolated. It was thus the ideal 'bounded but leaky' entity. Its major weakness was that it was contiguous with another, more warlike, peoples to the north, namely the Highland Scots. Fortunately for it, there were few highlanders and they usually only constituted a raiding nuisance. Nevertheless, the unification of the Crown in 1603 eased the threat of land-based invasion, and the events of 1715 and 1745 showed how relaxed the English had become about their land defences.

The advantages of not having large warlike land neighbours were immense. Basically it decreased the temptations to, and possibilities of, predation. Most obviously, a country like England never suffered the kind of levelling catastrophe afflicted by the Mongols, thus avoided a dismantling of its infrastructure. The devastating effects of being fought over by warring armies, which brought northern Italy and southern Germany toppling from their heights, never occurred. The Norman invasion, the Wars of the Roses or the English Civil War were as nothing compared to the normal experience of Continental countries.

Secondly, the balance of ruler and ruled was altered. A people could not be held to ransom by the threat of foreign invaders, whose continued hostility gave rulers a weapon with which to extract taxation and obedience from their people. They did not have to suffer from that most powerful tool of political absolutism, a standing army whose presence in England in the 1640s might well have ushered in a Continental-style absolutism. An unarmed populace can stand up to an unarmed ruler – or rather one who provides defence only through a strong navy.

A third consequence is on the temptation to external expansion. Such predation, which the English engaged in with some zeal in France and Scotland during the twelfth to fourteenth centuries, had a different nature from that which attracted the Hapsburgs or Louis XIV or had been a major factor in the destruction of the Roman democracy. In the case of Scotland the target was finite and the aim to eliminate a particular threat. In the case of France, the dynastic claims encouraged adventure, were clearly a luxury not a necessity. If the king wanted to adventure in that way, just as when he wanted to gain merit in the Crusades, he would have to pay for it – by acceding to the desires of the lords and commons. Instead of such wars leading to the increase of royal power, as taxation was raised and the people threatened, it was the warlike adventurer kings who conceded most to the people. Every concession tipped the balance and though temporary reprieves of a financial kind could be achieved, for instances by the confiscation of monastic lands, a ruler wishing to be adventurous or loved, like Elizabeth, was too firmly circumscribed to be able to do anything other than sell off the family silver (the demesne lands) and put her successors further at the mercy of a powerful third estate. At a certain point, probably after about 1650, the process was more or less irreversible. Each new gain in production now fed into increased production, rather than internal predation.

The other theory to account for England's peculiar history overlaps, but adds depth to this account. There were certain initial differences in England before the Norman invasion. Above all, being further from Rome, there was a more thorough sweeping away of all Roman vestiges. The Anglo-Saxons brought a legal and social system which placed an emphasis on the individual rather than the community or the clan and which emphasized individual property rights. In the absence of powerful kin groups, the 'feudal' structure of contractual ties to the political centre had developed in this early unified and homogeneous island kingdom. One language, one law and one currency were established by the ninth century over most of what is now England.

Then for a century after the Norman invasion this already different island world came close to the rest of north west European models of feudalism, except that there was no 'dissolution of the State'. A powerful monarchy maintained its control of the legal system, but delegated power downwards through the feudal links. Thus England reached a balance of centralization and de-centralization.

From a basic similarity by about 1200, 'Europe' increasingly gravitated towards other forms, with the reception of Roman Law and the growth of a blood nobility and a widespread peasantry. England, after incorporating some of the organizational features of Roman Law, rejected almost all the substantive content of the revived late Roman legal system. This meant that England's economic, social and political structure became more and more at variance with much of the continent. Montesquieu and Tocqueville guessed that England largely avoided the apparently inevitable tendency to 'caste' and 'political centralization'.

This unusual situation could not have happened if England had not been protected by the sea. It was possible to be a small replica of England, for a while, without a sea boundary, as in the Italian city-states or Holland, but to sustain such a balance over hundreds of years as a land-bounded nation was probably impossible. There was nothing inevitable about the process. Many islands have existed without any dramatic developments. It was only England's good fortune to be close to, but not too close, to a dynamic Continent which enabled it to achieve anything. That something which has changed the world happened, was remarkable, unpredictable, and the result of an unique combination of forces.

For what happened was that at a certain point the feedback loop had altered. That strange alchemy whereby military strength was fuelled by high level production, rather than feeding off it, had begun. The rich became the powerful, rather than the powerful becoming the rich. Extra production led to increased power and England was able to suck in the wealth of inventions and goods of not only Europe, but increasingly of America and Asia, to fuel its attempt to escape for the first time from the agrarian trap. It had, from another viewpoint, become the most powerful predator.

By the middle of the eighteenth century England had a legal system which was uniform, largely predictable, reasonably fair and well administered and which protected property and the person. It had a wealth of competing religious opinions and a residuary Calvinist ethic, but no absolutely dominating group despite the established Church. It had the most productive agriculture in the world. It had a fine craft tradition. It had a developed banking and insurance system, a huge and prosperous metropolis, and an obsession with making money. Its social structure consisted in the main of a large and unusually well educated and materially prosperous middling sort. It had some of the best inventors and scientists in the world.

Then within a century Britain had become the greatest Empire the world had ever known. Its language was very widely used. Its democratic political system had started to spread over the globe and its legal system was widely emulated. More generally its new science and technology, based on replacing human labour by carbon fuel, made it the workshop of the world.

The reverberations of this breakthrough were momentous enough, but they were made doubly so by an accidental replication of the central features. Elsewhere, in a large, lightly populated continent, to which the essentially 'modern' arrangements were transferred they soon performed the same magic, making it the richest country the world has ever known, namely the United States of America.

It was this version of the new Atlantis which attracted the attention of Adam Smith as the most dynamic part of the globe, and even more so Tocqueville, who saw in it the distillation of the central principles which had emerged in England, freed from a certain dross of aristocratic arrogance. Tocqueville's account of America provides an anatomy of how a system based on the premises of profit-seeking through commerce, religious freedom, political responsibility and the absence of instituted hierarchies of wealth worked. In its pure form it astonished him, and made it possible for him to discern the structural features which had been developing earlier in English history. Basically, the Americans had cut off the aristocratic and the labourer levels of the English model, but developed everything else to its limits – the legal system, associational tendencies, delegation of power and obsession with religious and personal liberty. This left them free, self-confident and obsessed with practical and profit-orientated activities to improve their lives. The balance between the family, the ritual world, the polity and the economy had been achieved. Smith's night watchman church and state were present, and an extreme form of individualism (self-love) was somehow combined with an extreme form of associational co-operation (the social) to form just that blend which would continue to lead to rapid growth for six generations. Its immense resources and the absence of the threat of war were added bonuses.

Yet America was a useful experiment in another sense for it again proved that what was essential was more than geographical good fortune. What was also crucial was the appropriate 'mores', that is the customs or configuration of a people. Tocqueville showed this through his comparison of the effects of an *ancien régime* in French Canada, which replicated the introverted, hierarchical, conservatism of the old France, with the dynamism and individualism of the area settled from England. It was part of a wider insight of several of our informants that wherever it was taken worked the same miracle, while the colonies of the Continental powers replicated the centralized, bureaucratic and hierarchical structures of their mother countries.

*

The optimistic, dynamic picture of progress does have a darker side, as our thinkers realized. Adam Smith had noted the disastrous mental and social effects of the very engines for growth. As people lost their aggressive, military, predatory desires, they lost their courage and honour. As they profited from the division of labour, they turned themselves into automata. Wealth was gained, meaning was lost.

By Tocqueville's time the miracle had been achieved and his main task was to warn of some of the dangers that would flow out of a new world of wealth, liberty and equality. He was one of the first to see how even these good things might, through over-ambition, become corrupted. Too much material wealth bred hedonism, restlessness, meaningless, an obsessive and never-ending striving for short-term profit. Too much liberty left individuals separate and alone, anomic and vulnerable. And too much equality reduced the character and political flexibility. Humankind was caught in a desperate competitive struggle to assert a little superiority. A person had no spiritual excuses if he or she fell behind. There was little solidarity with one's fellows. All of this, plus increasing hedonism, left mankind open to a new and more terrible form of despotism. This was not the old despotism of fear and repression of which Montesquieu had written, the despotism of the harem. It was the internal despotism, the annihilation of the individual will and conscience, of individual liberties, in a world that darkly anticipates a mixture of Orwell and Koestler.

It was a world which through one of those paradoxes which face us, awaits us just beyond all that is good. We now all applaud liberty, equality, wealth. Yet as they are carried to their extreme, they not only conflict with each other, but in themselves begin to erode those separations and dynamic tensions upon which they are based. This explains the paradox that the two thinkers who advocated the most extreme forms of equality and liberty, namely Rousseau and Marx, should have provided a blueprint for totalitarian systems which almost extinguished both. The French and Russian revolutions, founded on the noblest ideals, led to horrific dictatorships.

All this reminds us that everything lies in the relations and balances. Too little predation may be as bad as too much, too much production can enslave people as much as too little. Man's very greatness, as Pope realized, lay in his contradictions, in the tensions in his nature and his society. The riddle of his success is solved not by eliminating these tensions, but by confronting and rejoicing in them. The Puritan message of unending struggle within man and between man and nature is an attempt to grasp this message.

*

This, then, is the answer our four thinkers give us to the riddle of the emergence of the modern world. What are we to make of their theories? The argument explains how the differentiation of western Europe occurred. Furthermore, it answers another question, succinctly put by Ernest Gellner: 'if Teutons were free (some becoming corrupted in post 11th century continental Europe), then how did the Teutons differentiate themselves from other Indo-Europeans, some of whom most certainly were not friends of liberty (e.g. Persians)?'[1] The analysis shows both that we are not dealing with linguistic or other 'races' such as Indo-Europeans, but rather with a set of particular historical circumstances which include the influence of Greek civilization, of Christianity, of the decomposition of Roman civilization and of the geography and ecology of western Europe. It would clearly be a crude reduction of Montesquieu or Tocqueville's argument to suggest that they thought a 'Teutonic race' was somehow inherently bestowed with some genetic strain towards liberty.

Indeed one of the beauties of the theory is the way in which it distances itself from a rigid material determinism. Although all the informants took due account of the physical environment, stressing the importance of climate, geography, agricultural ecology, none of them saw this as over-ruling the more important aspect of what man makes of his environment. All of them put a balanced set of cultural, legal, social and political conditions at the forefront of their explanation. Nor did they fall into the vulgar Marxist trap of seeing everything determined 'in the last instance' by the relations of production, by the economic 'infrastructure'. They saw that it was impossible to separate the 'infra' and 'super' structure – everything was connected and legal structures, for instance, were not just a 'reflection' of something else, nor was religion just the 'opium of the people'. Both law and religion shaped the world through time, as much as they were shaped.

Another criticism might be that we have here some kind of timeless functionalism. With their mechanistic analogies, their talk of the whole and the parts, and the role of various institutions within a civilization, these informants might superficially be mistaken for precursors of the arid functionalism of certain twentieth century social thinkers. Yet this is also a mistaken charge. They did not see any particular point in time as a stable equilibrium created by a harmony of parts or interests. Rather they saw an endless movement and change in which conflicting pressures, at the individual and institutional level, would at any given time find a resolution. For although a civilization might get locked into an equilibrium of forces for a long time, as had China, it could at any time collapse as the jarring tides altered their strength. It was a dynamic, historical approach which saw change as normal and necessary, even if it was change over long periods.

Another trap they largely avoided was teleology. Although at times professing a mild Christian hope that things were progressing towards a better future, or, as in Tocqueville's case, a theory of the inevitable movement towards more equality, their central message was that there was no certain end. The 'victory' of the things they valued was not assured; all were deeply aware of the fragility of liberty, equality and even wealth. They knew enough about the history of the world to see that the previous great progressive civilizations had crashed or reached a sterile equilibrium. They hoped for a better world but were doubtful as to whether it would emerge. This is why we find that they can be labelled as neither pessimists or optimists about the future.

Furthermore, this is why it would be inaccurate to describe this as a 'Whig' view of history. At a superficial level it might be taken that the story they proposed of gradual growth of liberty, equality and wealth in England was the old Whig story. But each author took care to distance himself from the more insidious under-currents of such a philosophy. There was, first, nothing inevitable about the progress – indeed the normal course of history was for agrarian civilizations to lose their liberty, equality and even wealth. Thus there was no Hegelian 'spirit' of history which moved it inevitably in a certain direction, or which saw the 'end of history' in the early nineteenth century.[2] Nor is there the Marxist version, whereby the material conditions of life assure the 'normal' progress through all the modes of production to the final inevitable triumph of communism.

Indeed it was one of the disappointments of many of their admirers that they did not really seem to believe in 'progress'. Montesquieu believed in change, but could not assure his readers that the future would be 'better' than the present, or even that the present was better than all periods of the past. Smith saw a patchy growth in the wealth of nations, but could see no escape from the deep violence and strain of agrarian civilization. Tocqueville was often extremely gloomy about the future – both of his own France, and of the dangers that lay in the greatest exponent of the new potentials in America.

Thus they kept a strict balance. They recognized change, at times progress, at times a growing complexity. Yet they did not espouse an easy social Darwinian evolutionism. They did not believe in the survival of the 'fittest', unless 'fittest' meant tautologically the 'fittest to survive', which often meant, as with the Mongols, the most ruthless and warlike.

They thought that history could provide an arena where through conjectures and comparison one could test out theories. It warned of dangers, it suggested remedies, but it could not provide a sure guide to the future. All it did was to stress the degree of fragility of all that mankind holds good.

They were able to achieve this balance because, being philosophers as well as historians, all our writers had a deep appreciation of the double nature of man. At one level they accepted the Christian teaching of the fallen nature of man. They fully accepted that each person was ruled by 'vice', that is by egotistic and selfish desires – lust, greed, pride and the desire to dominate and humiliate. This was their basic premise. Yet they also saw that out of this there emerged all the glories that man had created, great art and literature and music, tenderness and altruism, complex civilizations of all sorts. Thus they neither fell into the naive utopianism of a Rousseau or Godwin or even, ultimately, a Marx, nor retreated into the dark pessimism of the early Malthus or Nietzsche. They were realists and pragmatists, who saw humans in the round, as bundles of contradiction, creative and curious, emulative and sycophantic.

The humans they described inter-acted with diverse eco-systems, created worlds of meaning and artifacts which were constantly changing. Their vision managed to combine both necessity and chance in the scheme of this change. They recognized the deep structural forces and tendencies which circumscribed the possibilities of each individual, the options open to him. In this sense they anticipated the best of the social sciences. Yet they also recognized the law of probabilities, the role of accident and 'chance', of the shape of Cleopatra's nose, and hence anticipate the best of the historical sciences and modern chaos theory. They avoided the sterile extremes; there is a surface of random chances, but at a deeper level there are also conditioning rules inscribed in the nature of man and the nature of the physical world.

Thus their methodology, a sophisticated use of the comparative method to provide tests of their theories and to generate questions, a deep interest in long historical periods to see how the balances altered over time, a constant attempt to look at the whole and the parts in their relations within that whole, was a very fruitful one. It did for the social and historical sciences what Bacon and Newton had done for the natural sciences. And in many ways it was based on that same delicate mixture of induction and deduction – looking back and forth between hypothesis and data, and never allowing one to dominate.

So, ultimately, we face the particular question – have they solved the riddle of the emergence of modern man? It is difficult to know how one would estimate this. They passed the practical test. The mixture of elements which they thought would generate wealth has indeed done so. It worked in the case of England and America – sensationally. When it was introduced to Japan, it had dramatic effects. Its total antithesis, the closed world of communism has collapsed in shambles and the liberal world of individualistic capitalism is spreading rapidly, only leaving pockets of an alternative, such as Islam. By these tests they were 'right'.

But what of their actual historical theory, in particular their account of the growing divergence between England and the Continent and the peculiarities of England? Such a theory was advanced by people whose access to detailed archival sources was fairly limited. Montesquieu, Smith and Tocqueville all worked with original records and thought deeply on the medieval period. But obviously a huge amount of microscopic research had been undertaken in the century and a half since they wrote.

This research at first seemed to fully confirm their conjectures. The work of the two greatest medievalists of the first half of the twentieth century, Maitland and Bloch gave magisterial support to their interpretations of English and French history.[3]

Then, as suggested above, a curious reversal occurred. A reaction, often inspired by vulgar Marxist interpretations, occurred which re-wrote the history of the world. This turned history into a uniform series of 'stages' or 'modes of production' through which all societies moved in sequence – ancient, medieval, early modern, modern; or pre-capitalist, proto-capitalist, capitalist; or pre-industrial, proto-industrial, industrial. English history, like the rest, was squeezed into this pattern. This was the theoretical world in which I was educated. The world was uniformly 'agrarian' until a certain part 'broke away' (for inexplicable reasons) to become 'industria.' Although such a view clearly appeals to many in the later twentieth century, this shows us more about ourselves than the past.

This is why it is so refreshing to go back to those who lived during the transformation itself and to see how much richer and complex their story is. Not only do they provide an answer, but they know the right questions to ask. They not only emphasize the miracle of what has happened, but through the tensions between an ideal model of the 'normal' tendency of agraria, and the accidental and chance deviations, show how the miracle, contravening the normal laws, may have happened.

I use the word 'may' advisedly. It is difficult to find many *factual* mistakes in their work; for instance, a thorough study of Montesquieu's historical writings comparing them to the most recent research by French historians shows that he made few serious mistakes and that almost all his findings have been substantiated by recent research.[4] Or again as Acton says of Tocqueville, he is 'always right' and subsequent historians of America have only found minor changes to his account need to be made.

Yet these thinkers themselves, with their modesty and balance would have been the last to assert a unique and hegemonic grasp of 'the truth'. They set out a possible picture, a conjecture of what has happened and why it happened. It is elegant, economic, self-consistent and allows predictions. It is open to refutation and to testing. It is 'science' in the true sense – that is to say it is a provisional approximation attempting to fit an immensely complex reality. It cannot be proved right, but only shown not yet to have been proved wrong. Until such a time as it is proved wrong, either by the discovery of new facts which have not hitherto refuted it, or by logical arguments which show that it is untenable, it continues to stand as the most comprehensive and plausible hypothesis to explain the riddle of the world. That is to say, it shows us how the transformation of humankind through time, and particularly the last 1000 years, has occurred and accounts for nearly final mastery of almost all the resources of this planet, including him or herself.

*

Yet even when the riddle appears to be resolved, like all good riddles it only leads on to another. This book has brought to the forefront the role of technological differences between civilizations and the importance of the scientific and industrial revolutions. These were the tools which made the exit permanent. Our thinkers explain how these tools could be so effective once they developed, but none of them seriously addresses the question of why such tools emerged at all, although all of them provide occasional insights into the process.

The Enlightenment analysis does provide a plausible account of some of the necessary conditions of the escape, yet something still eludes us. Although immensely important as preconditions, liberty, equality, a measure of individualism, the division of labour, are not sufficient to show how and why there has been such a massive transformation from agrarian to modern industrial society.

In a sense then, the Enlightenment thinkers have designed a beautiful car, showing how the parts work and interrelate. They even recognize the importance of the engine, growing wealth. Yet they do not provide us with any systematic analysis of how the car actually moves, for they do not look into the internal combustion mechanism of the growth of reliable knowledge expressed in science and artefacts.

What tends to be overlooked is the history of creative thought in relation to science and technology. The heart of the transformation they were trying to explain was ultimately industrialization. They explain how industrialization became possible. But they do not really attempt to explain why the scientific and industrial revolutions occurred in the West. We may wonder why this large gap is to be found in their otherwise comprehensive treatment.

One part of the explanation is, of course, the timing of their lives. Because the full dimensions of what has happened was not obvious even by the 1840s, the three earlier thinkers, Montesquieu, Smith and Tocqueville, have little to say about the development of science and technology, except at a general level, as in Smith's fascinating history of astronomy, or Tocqueville's acknowledgement of the power of industrialism. Smith would have been best placed to comment on these subjects, but he lived just too early to appreciate the likely impact of industrial power and technology.[5]

Only Ernest Gellner is late enough and aware enough of the power of scientific knowledge and technology to take some account of this factor. He did see these as the prime movers, the *deus ex machina* (or vice versa). Yet like the others he was more preoccupied with liberty, equality and the consequences of wealth. He therefore only gives a cursory account of the conditions of the development of science and technology.

A second reason may be given, which concerns the background and position of our thinkers. Montesquieu and Tocqueville, as aristocrats, show a certain disdain for those processes closely connected to the production of wealth. Gellner as an academic is likewise somewhat distanced from that world of manufacture and the application of science which helps to emphasize its importance. A tell-tale indication of this is the way in which these authors tend to lump 'trade' and 'manufacture', which are very different in their implications. Smith did not suffer from these blinkers, but his experience was too early to enable him to pursue the matter.

Whatever the reasons, the effect is clear. There is still a large gap in the explanation of how the transition to the modern world has occurred. Overcoming the Malthusian trap is part of the story, and I have tried to provide a theory to explain how that happened.[6] Partially overcoming political, religious and social predation is another part of the total picture and this book has provided a theory as to how that may have occurred. Yet there is a third trap which needs consideration. In order to complete the picture we need a thorough examination of the conditions which lead certain societies to go through an industrial revolution, and others an industrious one, some to go through a wisdom revolution and others through a knowledge (science) one. Or, put in another way, why did technological and scientific growth occur so spectacularly and rapidly in western Europe between about the twelfth and nineteenth centuries and why, during the same period, did it slow down, cease and even partially regress in other civilizations which had previously been far more 'advanced' than Europe?

Notes

1 The Riddle of the Modern World

1 I am grateful to Gerry Martin for both the form and content of the next three paragraphs.
2 See the table in Fukuyama, *End of History*, 49–50.
3 Mokyr, *Lever*, 303.
4 Tocqueville, *Ancien*, 6–7.
5 Fukuyama, *End of History*; for a brilliant earlier attack on such teleological approaches, see Berlin, 'Historical Inevitability' in *Four Essays*.

2 Baron de Montesquieu's Life and Vision

1 Shklar, *Montesquieu*, 3.
2 Althusser, *Montesquieu*, 18–19.
3 Sorel, *Montesquieu*, 30.
4 Tocqueville, *Ancien*, 89.
5 Montesquieu, *Spirit* I, 307, 308, 309, 312.
6 Quoted in Weber, *Protestant*, 45.
7 Cited in Collins, *Montesquieu*, 178.
8 Ibid. 178.
9 Ibid. 143; Montesquieu made a distinction between 'piety' and 'religion', which alone explains his apparently contradictory statements.
10 Ibid. 138.
11 Ibid. 178.
12 Ibid. 178–9.
13 Ibid. 181; for a more recent and detailed account of his English visit, see Shackleton, *Montesquieu*, ch. 6.
14 Ibid. 231.
15 Quoted in Conroy, *Montesquieu*, 126.
16 Ibid. 126.
17 Ibid. 127.
18 Montesquieu, *Spirit* I, 142.
19 Montesquieu, *Considerations*, 169.
20 Shackleton, *Montesquieu*, 168.
21 Ibid. 169; there are other particularly interesting discussions of Montesquieu's theories of causation in Sorel, *Montesquieu*, 205–7 and Durkheim, *Montesquieu*, 44–7.
22 Collingwood, *Idea of History*, 79, 97, 200.
23 Shackleton, *Montesquieu*, ch. 14, and Sorel, *Montesquieu*, 139–40 are particularly illuminating accounts of his theory of climate and causation.
24 Sorel, *Montesquieu*, 89.
25 Bury, *Progress*, 145, 148.

26 Richter, *Montesquieu*, 105.
27 Shklar, *Montesquieu*, 50; Fletcher, *Montesquieu*, 73 and Durkheim, *Montesquieu*, 57–9 argue similarly that Montesquieu had no sense of progressive time.
28 Shackleton, *Essays*, 37.
29 Shackleton, *Montesquieu*, 319.
30 Complete Works, quoted in Richter, 'Montesquieu', 472.
31 Durkheim, *Montesquieu*, 50–1.
32 Richter 'Montesquieu', 472.
33 Ibid. 475.
34 Ibid. 73.
35 The importance of his achievement is again well summarized by Durkheim, *Montesquieu*, 62.
36 Sorel, *Montesquieu*, 155.
37 Montesquieu, *Persian Letters*, no. 84.
38 Montesquieu, *Spirit* I, 1.
39 Ibid.
40 Sorel, *Montesquieu*, 86–7.
41 Durkheim, *Montesquieu*, 51.
42 Fletcher, *Montesquieu*, 74.
43 Shklar, *Montesquieu*, 101.
44 The best account of how Montesquieu founded modern functionalism and structuralism is in Durkheim, *Montesquieu*, 56–7, 63.
45 Shackleton, *Montesquieu*, 309–10.
46 Fletcher, *Montesquieu*, 72; my translation.
47 *Pensées*, quoted in Conroy, *Montesquieu*, 131.
48 Durkheim, *Montesquieu*, 52.
49 Ibid.
50 Durkheim, *Montesquieu*, 53; see also for Durkheim's criticisms, *ibid.* 54.
51 Shackleton, *Montesquieu*, 158.
52 Sorel, *Montesquieu*, 84–5.
53 Montesquieu, *Spirit* I, 85, 196, 233–4; II, 35.
54 Richter, *Montesquieu*, 84.
55 See Shackleton, *Essays*, 231ff; Hulliung, *Montesquieu*, 100ff.
56 Important passages occur in Montesquieu, *Spirit* I, 369, 297, 304, and especially 122–5.
57 Montesquieu to President Barbon, 2 Feb. 1742, translated from Starobonski, *Montesquieu*, 159.
58 Montesquieu, 'Dossier on Esprit', translated from Starobonski, *Montesquieu*, 160; Sorel, *Montesquieu*, 92, describes the danger of the ageing philosopher desperately trying to finish the great work.
59 Ibid. 52–3.
60 Montesquieu, 'Dossier on Laws', translated from Starobonski, *Montesquieu*, 182.

3 Liberty and Despotism

1 Montesquieu, *Spirit* I, 115.
2 Acton, *Life of Creighton* I, 372.

3 Montesquieu, *Spirit* I, 149.
4 Ibid. 109.
5 Shklar, *Montesquieu*, 38.
6 Ibid. 78.
7 Montesquieu, *Spirit* I, 161–2.
8 Shklar, *Montesquieu*, 81.
9 Ibid. 85.
10 Shackleton, *Montesquieu*, 272.
11 Ibid. 271.
12 Montesquieu, *Persian Letters*, no. 103.
13 Ibid. no. 106.
14 Ibid.
15 Montesquieu, *Spirit* I, 335.
16 Ibid. II, 260.
17 Ibid. I, 272.
18 Ibid.
19 Ibid. 273.
20 Ibid. 310.
21 Ibid. 217.
22 Ibid.
23 Ibid.
24 Ibid. 272.
25 Ibid. 257.
26 Montesquieu, *Considerations*, 94–5.
27 Ibid. 92.
28 Ibid. 91.
29 Ibid. 169.
30 Quoted in Richter, *Montesquieu*, 55.
31 Montesquieu, *Spirit* I, 369.
32 Ibid. 266.
33 Ibid. II, 29.
34 Ibid. I, 301.
35 Ibid. II, 64.
36 Ibid. 43.
37 Ibid. 31.
38 Ibid. I, 224.
39 Ibid. 273.
40 Ibid.
41 Ibid. 319.
42 Ibid. 271.
43 Ibid.
44 Ibid. 333.
45 Ibid.
46 Ibid. 310.
47 Ibid. 332.
48 Ibid. 316.
49 Ibid. 332.
50 Ibid. 316.
51 Ibid. 46.

52 Ibid. 316.
53 Ibid. 295.
54 Ibid.
55 Ibid. 393.
56 Ibid. 336.
57 Ibid. 329–30.
58 Ibid. 333.
59 Ibid. 332.
60 Ibid. II, 8.
61 Ibid. I, 298.
62 Ibid. 225.
63 Ibid. II, 8.
64 Ibid. I, 123.
65 Ibid. II, 6.
66 Ibid. I, 124.
67 Montesquieu, *Persian Letters*, no. 120; 'Tyen' is noted as 'the heaven of the Chinese'.
68 Montesquieu, *Spirit* I, 99.
69 Ibid. 124.
70 Ibid. 100.
71 Ibid. 124.
72 Ibid.
73 Ibid. 122.
74 Ibid.
75 Ibid. 269.
76 Ibid.
77 Ibid.
78 Ibid.
79 Durkheim, *Montesquieu*, 37–8.
80 Ibid. 38.
81 Quoted in Conroy, *Montesquieu*, 127.
82 Bloch, *French Rural History*, 35.

4 The Defence of Liberty

1 Montesquieu, *Persian Letters*, no. 105.
2 Ibid.
3 Sorel, *Montesquieu*, 126–7.
4 Montesquieu, *Spirit* I, 314.
5 Ibid. 312.
6 Ibid. 309.
7 Ibid. 308.
8 Ibid. 327.
9 Ibid.
10 Ibid. 315.
11 Ibid.
12 Ibid. 150.
13 Ibid.

14 Montesquieu, *Persian Letters*, 81.
15 Shklar, *Montesquieu*, 86.
16 Montesquieu, *Spirit* I, 151.
17 Ibid. 183.
18 Ibid.
19 Ibid. II, 78.
20 Ibid. I, 183–4.
21 Sorel, *Montesquieu*, 113.
22 Ibid.
23 Ibid.
24 Montesquieu, *Spirit* I, 321.
25 Ibid. 310.
26 Montesquieu, *Persian Letters*, no. 136.
27 Ibid. no. 118.
28 Ibid.
29 Ibid. no. 123.
30 Ibid. no. 86.
31 Montesquieu, *Spirit* I, 100.
32 Ibid. 314.
33 Ibid. 311–12.
34 Ibid. 310.
35 Ibid. 311.
36 Ibid. 57.
37 Ibid. 63.
38 Ibid. 396–7.
39 Montesquieu, *Persian Letters*, no. 107.
40 Montesquieu, *Considerations*, 171.
41 Montesquieu, *Spirit* I, 214.
42 Ibid. 310.
43 Ibid. 309–10.
44 Montesquieu, *Persian Letters*, no. 123.
45 Ibid. no. 107.
46 Quoted in Sorel, *Montesquieu*, 130.
47 Montesquieu, *Spirit* I, 310.
48 Ibid.
49 Ibid. 376.
50 Ibid. II, 73.
51 Ibid. I, 307.
52 Ibid. 303.
53 Ibid. 302.
54 Ibid. 152.
55 Ibid. 150.
56 Ibid. 152.
57 Shklar, *Montesquieu*, 81.
58 Montesquieu, *Spirit* I, 160.
59 See Shackleton, *Montesquieu*, 286, 298–301; Shackleton, *Essays*, 3ff, 7; Althusser, *Montesquieu*, 88; and for a defence of Montesquieu, Morgan, *Liberty of Thought*, esp. 13.
60 Montesquieu, *Spirit* I, 180.

61 Ibid. 160.
62 Ibid. II, 156.
63 Ibid. 110.
64 Richter, 'Montesquieu', 474.
65 Sorel, *Montesquieu*, 117.
66 Shackleton, *Montesquieu*, 279–80.
67 Montesquieu, *Spirit* II, 148.
68 Shklar, *Montesquieu*, 59.
69 Montesquieu, *Considerations*, 93.
70 Ibid. 93–4.
71 Ibid. 94.
72 Ibid. 47.
73 Montesquieu, *Spirit* I, 310.
74 Ibid. 321.
75 Ibid. 277.
76 Ibid. 240.
77 Ibid. 111.
78 Ibid. II, 172.
79 Ibid. 175.
80 Ibid. 95.
81 Ibid. I, 163.
82 Montesquieu, *Persian Letters*, no. 131.
83 Montesquieu, *Spirit* I, 376.
84 Ibid. 91.
85 Ibid. 376.
86 Montesquieu, *Persian Letters*, no. 131.
87 Montesquieu, *Spirit* I, 268.
88 Ibid. 232.
89 Ibid. II, 196.
90 Ibid. 171.
91 Ibid. I, 163.
92 Ibid. II, 171.
93 Ibid. I, 161.
94 Montesquieu, *Persian Letters*, no. 136.
95 Ibid. no. 93.
96 Ibid. no. 101.
97 Cox, *Montesquieu*, 21.
98 Ibid.
99 Ibid.
100 Collins, *Montesquieu*, 177.
101 Sorel, *Montesquieu*, 28–9.
102 Montesquieu, *Persian Letters*, no. 161.
103 Ibid.

5 Adam Smith's Life and Vision

1. On a printed postcard at the front of the King's College, Cambridge copy of Fay, *Adam Smith*.
2. Scott, *Smith*, 78.
3. Rae, *Smith*, 88-9.
4. Ibid. 91.
5. Ibid. 90-1.
6. Ibid. 89.
7. Scott, *Smith*, 81.
8. Ibid. 114.
9. Rae, *Smith*, 71.
10. Stewart, *Works* X, 42.
11. Meek, *Ignoble*, 128.
12. Ross, *Smith*, 83.
13. Stewart, *Works* X, 45
14. Rae, *Smith*, 197.
15. Ibid. 201.
16. Ross, *Smith*, 203.
17. Stewart, *Works* X, 43.
18. Rae, *Smith*, 264.
19. Ibid.
20. Ibid. 265-6.
21. Ross, *Smith*, 90.
22. Smith, *Philosophical*, 66.
23. Smith, *Moral*, 463-4.
24. See Ross, *Smith*, 216.
25. Quoted in Campbell and Skinner, *Smith*, 171.
26. Meek, *Ignoble*, 220-1.
27. Campbell and Skinner, *Smith*, 95.
28. Chamberlayne, 1737, quoted in Ross, *Smith*, 55.
29. Quoted in Campbell and Skinner, *Smith*, 92.
30. Smith, *Moral*, 288.
31. Ibid.
32. Campbell and Skinner, *Smith*, 67, quoting Stewart.
33. Ibid.
34. Cf. e.g. Smith, *Jurisprudence*, 201ff.
35. Meek, *Smith*, 29.
36. Ibid.; cf. Meek, *Ignoble Savage*, 35.
37. Skinner, *System*, 123ff.
38. See Skinner in Jones and Skinner, *Adam Smith Reviewed*, ch. 10.
39. Quoted in Stocking, 'Scotland', 66; see the whole article for an outline of Kames's thought.
40. See the interesting summary of the relations by Stern in his introduction to *Josiah Tucker*.
41. Ross, *Smith*, 54.
42. Leechman, 1755, quoted in Ross, *Smith*, 54.
43. See Mandeville, *Fable of the Bees*.
44. See Hume, *Essays*, particular the essays 'Of Commerce', 'Of the Rise and Progress of the Arts and Sciences', 'Of National Character'.

6 Growth and Stasis

1. Stewart, *Works* X, 68.
2. Ibid. 36.
3. Ibid. 60.
4. Smith, *Wealth* I, 17.
5. Smith, *Jurisprudence*, 352.
6. Smith, *Wealth* I, 26.
7. Smith, *Jurisprudence*, 334.
8. Ibid. 337.
9. Ibid.
10. Smith, *Jurisprudence*, 338.
11. Ibid.
12. Ibid. 339.
13. Smith. *Jurisprudence*, 346.
14. Ibid. 351.
15. Smith. *Jurisprudence*, 346–7.
16. Smith, *Philosophical*, 66.
17. Smith, *Rhetoric*, 102
18. Smith, *Jurisprudence*, 341.
19. Ibid. 339.
20. Smith, *Wealth* I, 15.
21. Ibid. 9.
22. Ibid. 11.
23. Smith, *Jurisprudence*, 350.
24. Smith, *Wealth* I, 13–14.
25. Smith, *Jurisprudence*, 346.
26. See Kindleberger and the comments by Briggs and Hartwell in Wilson and Skinner, *The Market* for a discussion of why Smith may have missed the implications of the new sources of power.
27. Smith, *Wealth* I, 26.
28. Smith, *Jurisprudence*, 356.
29. Ibid. 355.
30. Ibid. 538.
31. Ibid.
32. Ibid. 539.
33. Smith, *Rhetoric*, 137.
34. Ibid.
35. Cf. Meek, *Ignoble*, 238–9, who notes that Smith was setting up a 'tendency' which was very often not fulfilled.
36. Smith, *Jurisprudence*, 521.
37. Ibid. 522.
38. Ibid. 50.
39. Ibid. 223.
40. Ibid.
41. Ibid. 522.
42. Ibid.
43. Ibid.
44. Ibid.

45 Smith, *Wealth* I, 102.
46 Ibid. 223.
47 Ibid. 225.
48 Ibid. 395.
49 Ibid. 102.
50 Ibid. 100.
51 Ibid. 366.
52 Ibid. 80.
53 Ibid. 81.
54 Ibid. 80–1.
55 Ibid. II, 201.
56 Ibid. 202.
57 Ibid.
58 Ibid. I, 228.
59 Ibid.
60 Ibid. 229.
61 Ibid. 178.
62 Ibid. 179.
63 Ibid. 81.
64 Elvin, *Pattern*, 203.

7 Of Wealth and Liberty

1   Smith, *Wealth* I, 401.
2   Ibid.
3   Ibid. 432–3.
4   Ibid. 433.
5   Ibid. 423.
6   Ibid.
7   Ibid. 424.
8   Smith, *Jurisprudence*, 256.
9   Ibid.
10 Smith, *Wealth* I, 425–6.
11 Ibid. 426.
12 Ibid.
13 Smith, *Rhetoric*, 137.
14 Weber, *Cities*.
15 Smith, *Wealth* I, 402.
16 Ibid. 404.
17 Ibid. 405–6.
18 Stewart, *Works* X, 61.
19 Ibid. summarizing Smith's view.
20 Stewart, *Works* X, 61.
21 Smith, *Wealth* I, 276.
22 Ibid. 442.
23 Ibid.
24 Ibid. 443.
25 Ibid.

26 Ibid. 415.
27 Ibid. 418.
28 Ibid. 415.
29 Ibid. 413–14.
30 Ibid. II, 83–4.
31 Ibid. 98.
32 Ibid. I, 440.
33 Ibid. 409–10.
34 Ibid. II, 129.
35 Ibid. 350.
36 Ibid.
37 Ibid. 351.
38 Smith, *Jurisprudence*, 362.
39 Ibid. 529.
40 Ibid. 365.
41 Smith, *Wealth* I, 301.
42 Ibid.
43 Smith, *Jurisprudence*, 528.
44 Ibid. 524.
45 Ibid. 529.
46 Ibid. 39.
47 Ibid. 216.
48 Smith, *Moral*, 327.
49 Smith, *Jurisprudence*, 63.
50 Ibid.
51 Ibid. 65.
52 Ibid. 68.
53 See Goody, *Family*, ch. 6.
54 Smith, *Jurisprudence*, 68.
55 Ibid. 56.
56 Ibid. 49.
57 Ibid. 181.
58 Ibid. 182.
59 Ibid. 187.
60 Ibid. 189.
61 Smith, *Wealth* I, 367.
62 Ibid. II, 208.
63 Ibid. 208–9.
64 Ibid. 314.
65 Ibid. 315.
66 Ibid. 315–16.
67 Ibid. 317.
68 Ibid. I, 323.
69 Ibid. II, 325.
70 Ibid. 325–6.
71 Ibid. 328.
72 Ibid. 330.
73 Ibid. 331.
74 Ibid. 333.

8 From Predation to Production

1 Quoted in the *Oxford Dictionary of Quotations* (Oxford, 1950), 208b.
2 Smith, *Wealth* I, 435.
3 Ibid. 436.
4 Ibid. 437.
5 Ibid.
6 Ibid. 437–8.
7 Ibid. 438–9.
8 Ibid. 440.
9 Ibid.
10 Ibid. 441.
11 Ibid.
12 Ibid. II, 243–4.
13 Ibid. 243.
14 Ibid. 219–220.
15 Smith, *Jurisprudence*, 540–1.
16 Ibid. 540.
17 Ibid.
18 Smith, *Wealth* II, 220.
19 Smith, *Jurisprudence*, 230–1.
20 Ibid. 232.
21 Ibid. 233.
22 Ibid. 235.
23 Ibid. 243.
24 Ibid. 232.
25 Ibid. 192.
26 Ibid. 236.
27 Ibid. 234.
28 Smith, *Moral*, 218.
29 Smith, *Jurisprudence*, 220.
30 Ibid. 550.
31 Ibid.
32 Ibid.
33 Campbell and Skinner, *Smith*, 118.
34 Smith, *Jurisprudence*, 244.
35 Ibid. 250.
36 Ibid.
37 Ibid. 251.
38 Ibid. 252.
39 Ibid. 254.
40 Ibid. 255.
41 Ibid. 258.
42 Ibid. 264.
43 Ibid. 262.
44 Ibid. 264.
45 Ibid. 260.
46 Ibid.
47 Ibid. 262.

48 Ibid. 264.
49 Ibid. 266.
50 Ibid. 265.
51 Ibid.
52 Ibid. 266.
53 Ibid. 269.
54 Ibid. 271.
55 Ibid. 272.
56 Ibid. 274.
57 Ibid. 273.
58 Ibid. 274.
59 Ibid. 207.
60 Ibid. 271.
61 Ibid. 275.
62 Ibid. 280.
63 Ibid. 283.
64 Ibid. 284.
65 Ibid. 284–5.
66 Ibid. 272.
67 Ibid. 286–7.
68 Smith, *Moral*, 235.
69 Ibid. 197.
70 Roll, *Economic*, 152.
71 Stewart, *Works* X, 58.
72 Smith, *Wealth* II, 230.
73 Ibid.
74 Ibid. 230–1.
75 Wrigley, 'Two Kinds', 99–100.
76 Wrigley, *People*, 58.
77 Wrigley, 'Two Kinds', 115.
78 Smith, *Wealth* I, 108.
79 For further discussion of the rate of declining profit, see Wrigley, *People*, 29–32.
80 Wrigley, 'Two Kinds', 104.
81 Smith, *Wealth* I, 163.
82 Ibid. 90.
83 Ibid. 89.
84 Ibid.
85 Ibid. 88.
86 Wrigley, 'Two Kinds', 103.
87 For example the essays by R.L. Heilbronner and E.G. West in Skinner and Wilson, *Essays on Adam Smith*.
88 Campbell and Skinner, *Smith*, 121.
89 Smith, *Jurisprudence*, 539.
90 Ibid.
91 Ibid. 539–40.
92 Ibid. 541.

9 Alexis de Tocqueville's Life and Vision

1. For a further analysis of the deep contradictions in his personality, background and views see Boesche, *Tocqueville*, 16, 264–6.
2. Drescher, *Tocqueville*, 15.
3. Ibid.
4. Tocqueville, *Letters*, 100 (1835).
5. Quoted in Pierson, *Tocqueville*, 750.
6. Pierson, *Tocqueville in America*, 13–14.
7. Quoted in Jardin, *Tocqueville*, 377.
8. Tocqueville, *Letters*, 115–16 (1837).
9. Ibid. 113 (1836).
10. Ibid. 114 (1836).
11. Tocqueville, *Democracy* II, 913.
12. Ibid. 914.
13. Ibid.
14. Ibid.
15. Ibid. 916.
16. Laski, 'Tocqueville', 111.
17. Tocqueville, *Ancien*, 6–7.
18. For a good overview of Tocqueville's very considerable theoretical debt to Montesquieu, see Richter, 'Uses of Theory'.
19. Tocqueville, *Memoir* II, 358.
20. Ibid. I, 146,
21. Ibid.
22. Tocqueville, *Journeys*, xviii.
23. Ibid. 4.
24. Jardin, *Tocqueville*, 235.
25. Tocqueville, *Ancien*, 21.
26. Tocqueville, *Memoir* I, 359.
27. Quoted in Pierson, *Tocqueville*, 404.
28. Tocqueville, *Letters*, 58.
29. See, for further comments and examples, Pope, *Tocqueville*, 34ff; Schleifer, *America*, 71, 279.
30. Tocqueville, *Memoir* I, 304.
31. Ibid. 18.
32. Tocqueville, *Journeys*, xviii.
33. Tocqueville, *Memoir* II, 365.
34. Ibid.
35. See also Boesche in Nolla, *Liberty*, 180.
36. Quoted in Jardin, *Tocqueville*, 82.
37. Tocqueville, *Letters*, 355–6 (1857).
38. Tocqueville, *Democracy* II, 675.
39. Drescher, *Tocqueville*, 26.
40. Tocqueville, *Letters*, 99 (1835).
41. Tocqueville, *Journey to America*, 183.
42. Tocqueville, *Memoir* I, 193.
43. Tocqueville, *Democracy* I, 378–9.
44. Ibid. xl.

45 Ibid. 380.
46 Ibid. 381.
47 Ibid.
48 Drescher, *Tocqueville*, 30.
49 Ibid. 31.
50 Ibid. 33.
51 Tocqueville, *Journey to America*, 181.
52 Tocqueville, *Democracy* I, 35.
53 Ibid.
54 Ibid. 36.
55 Ibid. 345.
56 Tocqueville, 'Notes', 253.
57 Tocqueville, *Letters*, 60 (1831).
58 Mill, *Essays*, 257.
59 Drescher, *Tocqueville*, 215.
60 Ibid.
61 Pierson, *Tocqueville*, 757; see also Boesche, *Tocqueville*, 120 for a further discussion.
62 Tocqueville, *Democracy* I, xliii.
63 Jardin, *Tocqueville*, 274.
64 Quoted in Gargan, *Tocqueville*, 43.
65 Mill, *Essays*, 238.
66 Ibid. 216.
67 Ibid.
68 Ibid. 216–17.
69 Ibid. 181.

10 'America' as a Thought Experiment

1 Tocqueville, *Journey to America*, 183.
2 Drescher, *Tocqueville*, 215.
3 Pierson, *Tocqueville in America*, 747.
4 Tocqueville, *Democracy* I, 17.
5 Tocqueville, *Letters*, 95 (1834); Tocqueville and Beaumont were officially employed on a survey of the American prison (penitentiary) system.
6 Quoted in Tocqueville, *Democracy* I, p. liii.
7 Tocqueville, *Journey to America*, 177.
8 Tocqueville, *Democracy*, II, 1025.
9 Ibid. I, 17.
10 Pierson, *Tocqueville in America*, 70.
11 Tocqueville, *Democracy*, I, 287.
12 Ibid. 206–7.
13 Ibid. 207.
14 Tocqueville, *Letters*, 44 (1831).
15 Tocqueville, *Democracy* I, 285–6.
16 Ibid. 477.
17 Ibid. 483.
18 Ibid. II, 694.
19 Ibid. 796; see also Lerner's summary of these contradictions, ibid. I, p. c.

20 Ibid. I, 299.
21 Ibid. 298–9.
22 Tocqueville, *Journey to America*, 182; cf. Tocqueville, *Letters*, 44 (1831).
23 Tocqueville, *Journey to America*, 183.
24 Tocqueville, *Democracy* II, 693.
25 Ibid. 714; for Tocqueville's distaste for bourgeois society and particularly its obsession with wealth, see Boesche, *Tocqueville*, 85ff.
26 Quoted in Drescher, *Tocqueville*, 63–4.
27 Tocqueville, *Democracy* II, 796.
28 Ibid. 583.
29 Ibid. 805.
30 Ibid. 692.
31 Ibid. 593.
32 Ibid. I, 197.
33 Schleifer, *America*, 168.
34 Ibid. 283.
35 Ibid. 83.
36 Tocqueville, *Democracy* II, 711.
37 Ibid.
38 Ibid. 806.
39 Ibid. 712.
40 Ibid. I, 501.
41 Ibid. 350.
42 Ibid. II, 685.
43 Ibid. I, 352.
44 For example, Schleifer, *America*, 235–43.
45 Tocqueville, *Democracy* I, p. ciii.
46 Ibid. 351.
47 Ibid. 96.
48 Ibid. 96.
49 Tocqueville, *Journey to America*, 149.
50 Tocqueville, *Democracy* II, 826.
51 Ibid. 828.
52 Ibid. 621.
53 Ibid. 653.
54 Tocqueville, *Journey to America*, 183.
55 Tocqueville, *Democracy* I, 289–90.
56 Ibid. II, 696.
57 Ibid. 773–4.
58. Ibid. 701.
59 Ibid. I, 257.
60 Tocqueville, *Ancien*, 130–1.
61 Tocqueville, *Democracy* I, 301.
62 Tocqueville, *Letters*, 59 (1831).
63 Tocqueville, *Democracy* I, 297; Anderson, *Imagined Communities*, passim, esp. 5–7.
64 Tocqueville, *Democracy* I, 202–3.
65 Ibid. 203.
66 Ibid. 81.

67 Tocqueville, *Journey to America*, 179.
68 Tocqueville, *Democracy* II, 666.
69 Ibid. I, 232.
70 Ibid. II, 662.
71 Ibid. 663.
72 Weatherford, *Native Roots*, 183–4, and Mander, *Absence of the Sacred*, ch. 13, argue that the American settlers may have learnt some of the democratic and associational skills from the native Americans, for instance the famous Iroquois Confederacy, and that Tocqueville may have been half aware of this. I am grateful to Charles Ehrhart for these references.
73 Tocqueville, *Democracy* I, 89.
74 Ibid. 46–7.
75 Ibid. 87.
76 Ibid. 90.
77 For an excellent fuller account of this topic, see Goldstein, *Trial of Faith*.
78 Quoted in Boesche, *Tocqueville*, 110.
79 Tocqueville, *Ancien*, 165.
80 Tocqueville, *Letters*, 52 (1831).
81 Tocqueville, *Democracy* I, 369.
82 Tocqueville, *Letters*, 52 (1831).
83 Ibid. 49 (1831).
84 Tocqueville, *Democracy* I, 372.
85 Ibid. 365.
86 Ibid. 55.
87 Ibid.
88 Tocqueville, *Letters*, 50 (1831).
89 Ibid.
90 Ibid. 49 (1831).
91 Tocqueville, *Democracy* I, 357.
92 Ibid. II, 574.
93 Ibid. I, p. cii.
94 Ibid. II, 707.
95 Ibid. I, 5.
96 Ibid.
97 Ibid. 37.
98 Ibid. 17.
99 Ibid. 8.
100 Tocqueville, *Letters*, 56.
101 Tocqueville, *Democracy* II, 695.
102 Ibid. 720.
103 Ibid. I, 8.
104 Ibid. 9.
105 Tocqueville, *Journeys*, 9.
106 Ibid.
107 Tocqueville, *Democracy* I, 7.
108 Tocqueville, *Memoir* II, 282.
109 Tocqueville, *Democracy* I, 9.

110 Mill, *Essays*, 217.
111 Tocqueville, *Democracy* II, 756.
112 Ibid.
113 Ibid.
114 Ibid. 758.
115 Ibid. 759.
116 Ibid. 760.
117 Ibid. 758.
118 Ibid. 760.
119 Ibid. 763.
120 Ibid. 777.
121 Ibid.
122 Ibid. 779.
123 Ibid. 763.
124 Ibid.
125 Ibid. 767.
126 Ibid. 777–8.
127 Ibid. 780.
128 Ibid. 780–1.
129 Ibid. 766.
130 Ibid. 781.
131 There are numerous references to his views on slavery in the index to Tocqueville, *Democracy*.
132 For a good background account of the tragic impact of epidemics, trade goods, alcohol, firearms, mission schools and so on on Indian culture, see Axtell, *European and Indian*, ch. 9.
133 Tocqueville, *Letters*, 73 (1831).
134 Tocqueville, *Democracy* I, 393.
135 Ibid. 420.
136 Tocqueville, *Memoir* I, 202.
137 Tocqueville, *Journey to America*, 87.
138 Ibid. 160.
139 Tocqueville, *Democracy* II, 718.
140 Ibid. 719.
141 Ibid. I, 196.
142 Pierson, *Tocqueville in America*, 128.
143 Tocqueville, 'Notes', 270.
144 Ibid. 271.
145 Tocqueville, *Democracy* I, 37.
146 Ibid. II, 584–5.
147 Ibid. I, 36.
148 Ibid. 355–6.
149 Tocqueville, *Journey to America*, 177.
150 Tocqueville, *Democracy* I, 46.
151 Ibid. 37.
152 Ibid. 35.
153 Drescher, *Tocqueville*, 204; the quotation is from Tocqueville, *Ancien*, 53, and Tocqueville in fact is speaking of medieval Europe in general as the origin of the parish.

154 Tocqueville, *Democracy* II, 875–6.
155 Ibid. I, 45.

11 How the Modern World Emerged

1 Gargan, *Tocqueville*, 23; the essay is reprinted in Tocqueville, *Journeys*, 1–23.
2 Tocqueville, *Letters*, 82 (1833).
3 Quoted in Jardin, *Tocqueville*, 197.
4 Tocqueville, *Letters*, 83 (1833).
5 Drescher, *Tocqueville*, 37.
6 Tocqueville, *Letters*, 82 (1833).
7 Drescher, *Tocqueville*, 191.
8 Ibid. 123.
9 Ibid. 221.
10 Ibid.
11 Ibid. 103.
12 Ibid. 99.
13 Ibid.
14 Quoted Ibid. p. vii.
15 Ibid. 191.
16 Ibid. 221.
17 Ibid. 216.
18 Ibid. 204–5.
19 Tocqueville, *Letters*, 106–7; see also Drescher, *Tocqueville*, 73.
20 Jardin, *Tocqueville*, 339.
21 Ibid.
22 Tocqueville, *Ancien*, 89.
23 Tocqueville, *Journeys*, 44.
24 Drescher, *Tocqueville*, 110.
25 Tocqueville, *Memoir* II, 370.
26 Tocqueville, *Ancien*, 184.
27 Tocqueville, *Journeys*, 62.
28 Tocqueville, *Ancien*, 34.
29 Tocqueville, *Memoir* II, 376
30 Tocqueville, *Journeys*, 82.
31 Ibid. 105.
32 Ibid.
33 Ibid.
34 Ibid.
35 Ibid. 78.
36 Tocqueville, *Memoir* I, 230.
37 Tocqueville, *Democracy* II, 716.
38 Tocqueville, *Memoir* I, 159.
39 Tocqueville, *Journeys*, 104.
40 Ibid. 105.
41 Tocqueville, *Memoir* I, 186.
42 Ibid. 195–6.

43 Ibid. II, 54.
44 Ibid. 246.
45 Tocqueville, *Democracy* II, 714.
46 Ibid. 715.
47 Tocqueville, *Ancien*, 184–5.
48 Tocqueville, *Journeys*, 104.
49 Ibid. 60.
50 Tocqueville, *Democracy* II, 731.
51 Ibid. 732.
52 Tocqueville, *Memoir* II, 7.
53 Tocqueville, *Ancien*, 29.
54 Tocqueville, *Memoir* II, 7.
55 Ibid. 159.
56 Ibid. 147–8.
57 Tocqueville, *Journeys*, 13.
58 Ibid.
59 Tocqueville, *Democracy* I, 39.
60 Tocqueville, *Memoir* I, 311.
61 Ibid. 311–12.
62 Quoted in Drescher, *Tocqueville*, 127.
63 Tocqueville, *Letters* 108 (1836).
64 Tocqueville, *Memoir* II, 377.
65 Ibid.
66 Ibid.
67 Ibid. 378.
68 Ibid. 397.
69 Ibid. 353.
70 Ibid. 352–3.
71 Tocqueville, *Journeys*, 131.
72 Ibid. 47.
73 Tocqueville, *Memoir* I, 428.
74 Tocqueville, *Journeys*, 78.
75 Ibid. 56.
76 Ibid. 66.
77 Tocqueville, *Democracy* I, 331.
78 Tocqueville, 'Notes', 301.
79 Ibid. 302–3.
80 Drescher, *Tocqueville*, 91.
81 Ibid. 91–2.
82 Boesche, *Tocqueville*, 246.
83 Tocqueville, *Democracy* II, 1026–7.
84 Tocqueville, *Journeys*, 15.
85 Ibid. 107.
86 Ibid. 106.
87 Tocqueville, *Ancien*, 184.
88 Tocqueville, *Journeys*, 74.
89 Tocqueville, *Democracy* I, 45.
90 Tocqueville, *Journeys*, 2.
91 Ibid. 3.

92 Ibid.
93 Ibid. 2–3.
94 Tocqueville, *Memoir* II, 223–4.
95 Tocqueville, *Ancien*, 18–19.
96 Ibid. 92.
97 Ibid. 105.
98 Tocqueville, 'Notes', 242.
99 Tocqueville, *Memoir* I, 428.
100 Tocqueville, 'Notes', 240.
101 Ibid. 241–2.
102 Tocqueville, *Ancien*, 21.
103 Ibid. 89.
104 Ibid. 95.
105 Ibid. 93.
106 Ibid. 105.
107 Ibid. 92.
108 Ibid. 96.
109 Ibid. 89–90.
110 Ibid; in fact, the term *roturier*, meaning someone who held by an annual rent, had never existed in England.
111 Ibid. 90.
112 Ibid. 95.
113 Tocqueville, *Journeys*, 5.
114 Tocqueville, *Ancien*, 23.
115 Tocqueville, *Letters*, 355 (1851).
116 Tocqueville, *Memoir* II, 226.
117 Tocqueville, *Ancien*, 86.
118 Tocqueville, *Journeys*, 79.
119 Ibid. 78.
120 Tocqueville, *Recollections*, 84.
121 Tocqueville, *Democracy* II, 695; cf. also I, 243–4 on Pascal.
122 Ibid. II, 694.
123 Tocqueville, *Journeys*, 84.
124 Tocqueville, *Memoir* II, 160.
125 Tocqueville, *Democracy* II, 841.
126 Ibid. I, 209.
127 Tocqueville, Democracy, 378.
128 Ibid. I, 209.
129 Ibid.
130 Ibid.
131 Ibid. 210.
132 Ibid. 378.
133 Ibid. 345.
134 Ibid. II, 1026.
135 Ibid. I, 207.
136 Ibid. II, 879.
137 Ibid. II, 842.
138 Ibid. 879; a slight exaggeration, as the long peace in China shows.
139 Ibid. 842.

140 Ibid. I, 276.
141 Ibid. 198.
142 Ibid. 196.
143 Ibid.
144 Ibid. 197.
145 Ibid. II, 842.
146 Ibid. 198.
147 Ibid. 505.
148 Ibid. II, 856.

12 Liberty, Wealth and Equality

1 Tocqueville, *Journeys*, 150.
2 Tocqueville, *Memoir* I, 390.
3 Ibid. 305.
4 Ibid. 398.
5 Tocqueville, *Journeys*, 97.
6 Ibid. 96.
7 Ibid. 106.
8 Tocqueville, *Democracy* I, 52–3; Tocqueville, *Journeys*, 30.
9 Drescher, *Tocqueville*, 127.
10 Tocqueville, *Democracy* II, 649.
11 Ibid. 895.
12 Ibid. 882.
13 Ibid. 892.
14 Ibid. 893.
15 Ibid. 888.
16 Tocqueville, *Ancien*, 65.
17 Ibid. 64.
18 Tocqueville, *Democracy* II, 595.
19 Ibid. 881.
20 Ibid. I, 296.
21 Ibid. II, 899.
22 Tocqueville, quoted in Hearnshaw, 'Tocqueville', 109.
23 Tocqueville, *Democracy* II, 903.
24 Drescher, *Tocqueville*, 191.
25 Tocqueville, *Journey to America*, 51.
26 Tocqueville, *Ancien*, 178.
27 Ibid. 177.
28 Tocqueville, *Letters*, 66 (1831).
29 Tocqueville, *Democracy* II, 911.
30 Ibid. 916.
31 Tocqueville, *European Revolution*, 309–310.
32 Tocqueville, *Journeys*, 52.
33 Tocqueville, *Memoir* I, 311.
34 Ibid. II 53.
35 Tocqueville, *Democracy* II, 903.
36 Ibid. I, 117.

37 Quoted in Jardin, *Tocqueville* 273.
38 Tocqueville, *Memoir* I, 330.
39 Tocqueville, *Democracy* II, 657.
40 Tocqueville, *Ancien*, p. xv.
41 Ibid. 172.
42 Ibid. 102.
43 Ibid. 103.
44 Tocqueville, *Democracy* II, 652.
45 Drescher, *Tocqueville*, 197.
46 Tocqueville, *Democracy* II, 836.
47 Ibid. I, 79.
48 Ibid. II, 680.
49 Ibid. 653.
50 Ibid. 654.
51 Ibid. 836.
52 Ibid. 783.
53 Tocqueville, *Memoir* I, 10.
54 Tocqueville, *Recollections*, 33.
55 Tocqueville, *Democracy* II, 898.
56 Ibid.
57 Ibid. I, 315.
58 Boesche, *Tocqueville*, 258–9.
59 Tocqueville, *Ancien*, 125.
60 Ibid. 144.
61 Tocqueville, *Democracy* I, 67.
62 Ibid. 311.
63 Ibid. 66.
64 Ibid. 495.
65 Tocqueville, *Journeys*, 4.
66 Tocqueville, *Recollections*, 72.
67 Tocqueville, *Journeys*, 83.
68 Ibid. 11.
69 Tocqueville, *Memoir* II, 355.
70 Ibid. I, 211.
71 Ibid. 35.
72 Tocqueville, *Journeys*, 98.
73 Ibid. 98–9.
74 Quoted in Drescher, *Tocqueville*, 89.
75 Tocqueville, *Journeys*, 98–9; final dots are Tocqueville's.
76 Ibid. 85.
77 Ibid. 45.
78 Tocqueville, *Memoir* II, 226.
79 Tocqueville, *Journeys*, 75.
80 Ibid. 4.
81 Ibid.
82 Tocqueville, *Democracy* I, 87.
83 Ibid. 85.
84 Ibid. 89.
85 Ibid. 106.

86 Ibid. II, 885.
87 Ibid. I, 74.
88 Drescher, *Tocqueville*, 144.
89 Tocqueville, *Democracy* II, 596.
90 Ibid. 595.
91 Ibid. 596.
92 Ibid. I, 111, note 50.
93 Ibid. II, 821.
94 Ibid. 816–17.
95 Tocqueville, *Democracy*, I, 106.
96 Ibid. 107.
97 Tocqueville, *Journeys*, 162.
98 Tocqueville, *Democracy* I, 234.
99 Ibid. II, 906.
100 Ibid. 906.
101 Ibid. 667.
102 Tocqueville, *Journey to America*, 174.
103 Tocqueville, *Democracy* I, 336.
104 Ibid. 341.
105 Ibid. 339.
106 Ibid.
107 Ibid. II, 907.
108 Ibid. I, 91.
109 Quoted in Drescher, *Tocqueville*, 83.
110 Tocqueville, *Democracy* II, 888.
111 Ibid. I, 104.
112 Ibid. 50.
113 Ibid. 94.
114 Tocqueville, *Memoir* II, 238.
115 Ibid. 397.
116 Tocqueville, *Journeys*, 105.
117 Ibid. 105–6.
118 Tocqueville, *Ancien*, 163.
119 Tocqueville, *Memoir* II, 398.
120 Tocqueville, *Democracy* I, 360.
121 Ibid.
122 Ibid. 364.
123 Ibid. 359.
124 Ibid. II, 569.
125 Ibid. I, 115.
126 Tocqueville, *Journey to America*, 31.
127 Tocqueville, *Democracy* I, 357.
128 Tocqueville, *Journey to America*, 150.
129 Tocqueville, *Democracy*, I, 55.
130 Tocqueville, *Letters*, 356 (1857).
131 Tocqueville, *Democracy* I, 368.
132 Tocqueville, *European Revolution*, 206.
133 Tocqueville, *Journeys*, 74.
134 Ibid.

135 Ibid. 75.
136 Ibid.
137 Ibid. 74–5.
138 Ibid. 74.
139 Tocqueville, *Ancien*, 83.
140 Tocqueville, *Democracy* I, 238.
141 Ibid. 236.
142 Ibid. II, 883.
143 Ibid. I, 236.

13 Ernest Gellner and the Conditions of the Exit

1 'Conditions of the Exit' is a section title in Gellner, *Plough*.
2 Musil in *Gellner*, 42–3.
3 Gellner, *Muslim Society*, 1.
4 Gellner, *Liberty*, 29.
5 Ibid. 137.
6 Ibid. 210.
7 Gellner, *Plough*, 42.
8 For example, see Gellner, *Legitimation*, 169, 173.
9 Popper, *Open Society*, and Kuhn, *Scientific Revolutions*.
10 Gellner, *Legitimation*, 182.
11 Gellner, *Muslim*, 36–7.
12 Gellner, *Spectacles*, 285.
13 Gellner, *Plough*, 158.
14 Gellner, *Muslim*, 6.
15 Gellner, *Plough*, 277.
16 Gellner, *Liberty*, 7.
17 Ibid. 10.
18 Ibid. 55.
19 Ibid. 212.
20 Ibid. 80.
21 Ibid. 193.
22 Ibid. 141.
23 Gellner, *Anthropology*, 35.
24 Ibid. 33–4.
25 Ibid. 161.
26 Ibid.
27 Gellner, *Liberty*, 8.
28 Ibid. 32–3.
29 Ibid. 169.
30 Ibid. 140.
31 Ibid. 109.
32 Gellner, *Plough*, 204.
33 Gellner, *Anthropology*, 36.
34 Ibid. 39.
35 Gellner, *Plough*, 132.
36 Ibid. 122.

37 Gellner, *Devil*, 117–8.
38 Gellner, *Anthropology*, 37.
39 Gellner, *Liberty*, 45.
40 Gellner, *Anthropology*, 58.
41 He draws heavily and explicitly, though critically, on the ideas of Adam Smith and other Enlightenment authors here. For example, see his interesting long discussion of Adam Smith's theory concerning the reasons for the decline of violence, the effect of cities etc. in Gellner, *Culture*, 19–27.
42 Gellner, *Anthropology*, 168. It is obvious that both the word 'thug' and 'trader' are considerable oversimplifications. Many 'predators' are far from naked thugs, and as important as trade is production of wealth, manufacture, which Gellner tends to lump with 'trade'.
43 Gellner, *Liberty*, 169.
44 Ibid. 73.
45 Gellner, *Anthropology*, 167.
46 Gellner, *Liberty*, 68.
47 Gellner, *Anthropology*, 168.
48 Gellner, *Liberty*, 206. Here again Gellner lumps trade and manufacture together too simply. It is also not clear, in the light of 'corporate raiding' and other devices in the stock market that predation has indeed been subordinated.
49 Gellner, *Anthropology*, 131.
50 Gellner, *Liberty*, 79.
51 Ibid. 73.
52 Gellner, *Anthropology*, 168.
53 Gellner, *Liberty*, 33.
54 Gellner, *Anthropology*, 169.
55 Gellner, *Liberty*, 74.
56 Gellner, *Anthropology*, 167.
57 Gellner, *Liberty*, 74.
58 Ibid. 31.
59 Gellner, *Nations*, 5, 114.
60 In an interview with John Davis, May 1990, published in *Current Anthropology*, vol. 32, no. 1, Feb. 1991.
61 In Hall and Jarvie, *Gellner*, 663.
62 Gellner, *Liberty*, 103.
63 Ibid. 99–100.
64 Ibid. 100.
65 Ibid.
66 Gellner, *Plough*, 158.
67 Gellner, *Liberty*, 148.
68 Maitland, *History* II, 232–3.
69 Gellner, *Nations*, 19.
70 Gellner, *Plough*, 222, 112.
71 Ibid. 277, 199.
72 Hall and Jarvie, *Gellner*, 662.
73 Hall and Jarvie, Social, 663.
74 Ibid. 666.

14 The Riddle Resolved?

1 *Gellner*, 664.
2 See Fukuyama, *End of History*.
3 I hope to give a detailed account of Maitland's account in a separate work.
4 See Cox, *Montesquieu*, 21.
5 See the essay by C.P. Kindleberger in Wilson and Skinner, *The Market*.
6 Macfarlane, *Savage Wars of Peace*.

Bibliography

The bibliography includes all works referred to in the text, except for those by the four main thinkers, Gellner, Montesquieu, Smith and Tocqueville. Their works are listed at the front of the book. All books are published in London, unless otherwise indicated.

The following abbreviations have been used.

ed.	edited or editor
edn	edition
eds.	editors
Jnl.	Journal
n.d.	no date
tr.	translated by
Univ.	University

Althusser, Louis, *Politics and History; Montesquieu, Rousseau, Hegel and Marx*, tr. Ben Brewster, 1972.
Anderson, Benedict, *Imagined Communities*, 1991.
Axtell, James, *The European and the Indian: Essays in the Ethnohistory of Colonial North America*, Oxford, 1981.
Berlin, Isaiah, *Four Essays on Liberty* (Oxford, 1969).
Boesche, Roger, *The Strange Liberalism of de Tocqueville*, Cornell, 1987.
Boesche, Roger, 'Why Did Tocqueville Think a Successful Revolution was Impossible' in Nolla, Eduardo (ed.), *Liberty, Equality, Democracy*, New York, 1995.
Bury, J.B., *The Idea of Progress*, 1921.
Campbell, R.H. and Skinner, A.S., *Adam Smith*, 1982.
Chamberlayne, E., *The Present State of England*, 19th impression, 1700.
Collingwood, R.G., *The Idea of History*, Oxford, 1961.
Collins, J.C., Churton, *Voltaire, Montesquieu and Rousseau in England*, 1908.
Conroy, Peter, *Montesquieu Revisited*, New York, 1992.
Cox, Iris, *Montesquieu and the History of French Laws*, Oxford, 1983.
Drescher, Seymour, *Tocqueville and England*, Cambridge, Mass., 1964.
Durkheim, Emile, *Montesquieu and Rousseau: Forerunners of Sociology*, Ann Arbor, MI, 1965.
Elvin, Mark, *The Pattern of the Chinese Past*, 1973.
Evans-Pritchard, E.E., *A History of Anthropological Thought*, 1981.
Farrer, J.A., *Adam Smith*, Altrincham, 1988.
Fay, C.R., *The World of Adam Smith*, Cambridge, 1960.
Fitzgibbons, Athol, *Adam Smith's System of Liberty, Wealth and Virtue: The Moral and Political Foundations of the Wealth of Nations*, Oxford, 1995.
Fletcher, F.T.H., *Montesquieu and English Politics (1750–1800)*, 1939.
Fukuyama, Francis, *The End of History and the Last Man*, 1992.

Gargan, Edward T., *De Tocqueville*, Cambridge, 1965.
Goldstein, Doris S., *Trial of Faith; Religion and Politics in Tocqueville's Thought*, Oxford, 1975.
Goody, Jack, *The Development of the Family and Marriage in Europe*, Cambridge, 1983.
Hay, Denys, *Annalists and Historians: Western Historiography from the Eighth to the Eighteenth Centuries*, 1977.
Hulliung, Mark, *Montesquieu and the Old Regime*, Berkeley, Calif., 1976.
Hume, David, *Essays, Literary, Moral and Political*, n.d. (*c*.1870); Ward, Lock & Tyler reprint of 2 vols. 8vo edn.
Jardin, Andre, *Tocqueville: A Biography*, tr. Lydia Davis and Robert Hemenway, 1988.
Jones, Peter, and Skinner, Andrew S. (eds), *Adam Smith Reviewed*, Edinburgh, 1992.
Kames, Lord, *Sketches of the History of Man*, Basil (Basle), 1796.
Kuhn, Thomas, *The Structure of Scientific Revolutions*, Chicago, 1975.
Laski, Harold, 'Alexis de Tocqueville and Democracy', in *The Social and Political Ideas of Some Representative Thinkers of the Victorian Age*, ed. F.J.C. Hearnshaw, 1933.
Lux, Kenneth, *Adam Smith's Mistake*, 1990.
Macfarlane, Alan, *The Origins of English Individualism*, Oxford, 1978.
Macfarlane, Alan, *The Culture of Capitalism*, Oxford, 1987.
Macfarlane, Alan, 'Ernest Gellner and the Escape to Modernity', in *Transition to Modernity*, ed. John Hall and Ian Jarvie, Cambridge, 1992.
Macfarlane, Alan, *The Savage Wars of Peace: England, Japan and the Malthusian Trap*, Oxford, 1997.
Maitland, F.W. and Pollock, F., *History of English Law before the Time of Edward I*, with Sir F. Pollock, originally published in 1895, 2nd edn., Cambridge, 1923; preface by S.F.C. Milsom to the reprint of 2nd edn., Cambridge, 1968.
Mander, Jerry, *In the Absence of the Sacred: The Failure of Technology and the Survival of the Indian Nations*, San Francisco, 1988.
Mandeville, Bernard, *The Fable of the Bees*, ed. F.B. Kaye, 2 vols, Oxford, 1924.
Marshall, Alfred, *Principles of Economics*, 5th edn, 1907.
Marx, Karl, *Capital*, vol. III, London, 1954.
Meek, Ronald L., *Social Science and the Ignoble Savage*, Cambridge, 1976.
Meek, Ronald L., *Smith, Marx, & After*, 1977.
Mill, John Stuart, *Essays on Politics and Culture*, New York, 1963 (Anchor books edn).
Morgan, Charles, *The Liberty of Thought and the Separation of Powers*, Oxford, 1948.
Mokyr, Joel, *The Lever of Riches* (Oxford, 1992).
Nisbet, Ronald, *The Sociological Tradition*, 1967.
Oxford Dictionary of Quotations, Oxford, 1950.
Pierson, George Wilson, *Tocqueville and Beaumont in America*, New York, 1938.
Pope, Whitney, *Alexis de Tocqueville: His Social and Political Theory*, 1986.
Popper, Karl, *The Open Society and Its Enemies*, 1966.
Rae, John, *Life of Adam Smith* (1895), New York, 1965.

Richter, Melvin, 'Montesquieu', *Encyclopedia of the Social Sciences*, 2nd edn, 1968.
Richter, Melvin, 'The Uses of Theory: Tocqueville's Adaptation of Montesquieu', in *Essays in Theory and History*, ed. Melvin Richter, Cambridge, Mass., 1970.
Richter, Melvin, *The Political Theory of Montesquieu*, Cambridge, 1977.
Roll, Eric, *A History of Economic Theory*, 1961.
Ross, Ian Simpson, *The Life of Adam Smith*, Oxford, 1995.
Rostow, Walt W., *The Stages of Economic Growth*, Cambridge, 1962.
Schleifer, James T., *The Making of Tocqueville's Democracy in America*, 1980.
Scott, William Robert, *Adam Smith as Student and Professor*, Glasgow, 1937.
Shackleton, Robert, *Montesquieu: A Critical Biography*, Oxford, 1961.
Shackleton, Robert (ed.), *Essays on Montesquieu and Enlightenment*, 1988.
Shklar, Judith N., *Montesquieu*, Oxford, 1987.
Siedentop, Larry, *Tocqueville*, Oxford, 1994.
Skinner, Andrew S., *A System of Social Science: Papers Relating to Adam Smith*, Oxford, 1979.
Skinner, Andrew S., and Wilson, Thomas (eds), *Essays on Adam Smith*, Oxford, 1975.
Sorel, Albert, *Montesquieu*, 1887 (reissue of 1969).
Starobinski, Jean, *Montesquieu par lui-même*, Paris, 1979.
Stern, Jeffrey (ed.), *The Collected Works of Josiah Tucker*, 1993.
Stewart, Dugald, 'Biographical Memoirs of Adam Smith, LL.D.', in *Collected Works*, X, Edinburgh, 1858.
Stocking, George W., 'Scotland as the Model of Mankind: Lord Kames' Philosophical View of Civilization', in *Toward a Science of Man: Essays in the History of Anthropology*, ed. Timothy Thoresen, The Hague, 1975.
Weatherford, Jack, *Native Roots: How the Indians Enriched America*, New York, 1991.
Weber, Max, *The Protestant Ethic and the Spirit of Capitalism*, 1930, tr. Talcott Parsons.
Wilson, Thomas, and Skinner, Andrew S., *The Market and the State: Essays in Honour of Adam Smith*, Oxford, 1976.
Wrigley, E.A., 'Two Kinds of Capitalism, Two Kinds of Growth', *LSE Quarterly* 2:2, Summer 1988.
Wrigley, E.A., *Peoples, Cities and Wealth: The Transformation of Traditional Society*, Oxford, 1992.
Zetterbaum, Marvin, *Tocqueville and the Problem of Democracy*, 1967.

Index

Acton on power 29
agraria 262–4, 267–8, 271
agriculture 41–3, 52, 144
America 158, 161–3, 167–92, 217–18, 236, 285–6
American colonies 80–1
American Indians 188, 199
Ancien Régime 154, 157, 163–4, 190, 194, 195, 196, 209–10
  see also Tocqueville, Alexis de
Asia 26–7, 35–6
associations 178–9, 206–7, 244–6

Beaumont, Gustave de 153, 154, 171, 193, 206
Bolingbroke 88
Bouring, Dr 235

Caesar 62, 131
Canada 162
caste 210–16
Catholicism *see* religion
cause and effect 17–19
centralization/decentralization 176–9, 190, 206–7, 218–19, 223–4, 233–9, 284
Chesterfield, Earl of 67
China 42–3, 56–7, 102–6, 237–8, 270, 275–6
Christianity *see* religion
Civil Society 254–5, 264–5
civility 97–8
civilization, four-stage theory of 85
climate and resources 37–8, 40
Cochrane, Andrew 75
Collins, J.C. Churton 16, 17
commerce 37–41, 97–8, 107–8, 128, 222
Conditions of Liberty 254
  see also Gellner, Ernest

de Beaumont, Gustave 153, 154
de Montesquieu, Baron *see* Montesquieu, Baron de
de Tocqueville, Alexis *see* Tocqueville, Alexis de
decentralization/centralization 176–9, 190, 206–7, 218–19, 223–4, 233–9
deductive/inductive thought 24–6
democracy and equality 164–5, 183–4
Democracy in America 153–4, 156–7, 163, 164–5, 167–8, 184, 190, 225, 226, 227–8
  see also Tocqueville, Alexis de
Descartes 258
despotism 29–46, 224, 227, 230–2, 242, 286–7
development of mankind 2–8
division of labour 88, 94–7, 127–9, 147, 189
  see also invention
Drescher, Seymour 162–3, 164–5, 193, 194–5, 206–207
Durkheim, Emile 44–5, 255, 258

Earl of Chesterfield 67
East Asia, Montesquieu's writing on 26–7
Elizabeth I 135–6
England 15–17, 47–61, 113–15, 132–40, 158–9, 193–220, 234–9, 280–5
equality 167–92, 193–220
  and democracy 164–5
  liberty, wealth and 221–47, 286–7
Essay on Man 87, 88

Ferguson, Adam 255, 260
feudalism 208–9, 254, 265–6, 277–8
Fletcher, F.T.H. 24
foreign trade 37–41, 111–12
Foulis, Robert 77
four-stage theory of civilization 85
France 66–7, 195, 197–8, 213–15, 235–7

324

freedom of the press 239
French Revolution 157

Gargan 193
Gellner, Ernest 251–68, 287, 293
Germanic societies 62–5
growth and stasis 90–106
Guizot 160–1, 193

Holland 144, 279
Home, Henry, Lord Kames 87
Hume, David 78, 80, 81, 88–9, 107, 121, 252, 258–9
Hutcheson, Frances 87

'ideal type' analysis 22
individualism 228–30
inductive/deductive thought 24–6
interpersonal relations 185–7
invention 91–4
 see also division of labour
invisible hand theory 88
Islam 252

Johnson, Samuel 123
juries see laws
justices of the peace see laws

Kames, Lord, Henry Home 87

L'Ancien Régime see Ancien Régime
Laski, Harold 157
laws 23–4, 48, 138–40, 179, 202–6, 210–12, 239–41
Le Commerce 154
Lectures on Jurisprudence 79, 132
 see also Smith, Adam
Lerner 162, 165, 173, 182
Lettres Persanes 13, 15, 23, 68
 see also Montesquieu, Baron de
Lewis, Lady Theresa 203
liberty, defence of 47–69
liberty and despotism 29–46
liberty and wealth 107–22
liberty, wealth and equality 221–47, 286–7
Lingard 193
Louis XIV 15, 234, 236–7
Luther, Martin 18

Maclaurin, Colin 83
Maine, Sir Henry 265
Maitland, F.W. 265
Malinowski, Bronislaw 251
Mandeville, Bernard 88, 221, 276
mankind's development 2–8
mechanical analogies as used by Smith 81–3
Meek, Ronald L 77, 82, 85
military power 31–5, 51, 53, 142–3
 see also violence, ending of
Mill, J.S. 164, 165–6, 168, 184
Montesquieu, Baron de 13–28, 81, 85–6, 87, 161, 162, 218, 219, 233, 234, 241, 242, 243, 265, 277, 282, 289, 291, 293
 defence of liberty 47–69
 liberty and despotism 29–46
Musil, Jiri 251–2
Muslim Society 251
 see also Gellner, Ernest

Newton, Sir Isaac 81–2, 83

parent–child relations 185–7
parliamentary systems 48, 58, 137–8
Pensées 18, 45–6
 see also Montesquieu, Baron de
Persian Letters see Lettres Persanes
Pierson, George Wilson 154–5, 165, 167
political liberty 49–50, 58–9, 202–4, 225
political unit size 43–5
Pope, Alexander 1, 87, 88, 276
population growth 144–5
power 29–30
predation to production 123–49, 259–62, 270–4, 283
private vice, public benefit 88
production 123–49
property law 113–14
Protestantism see religion

Rae, John 77, 78, 80–1
Recollections 154
 see also Tocqueville, Alexis de

religion 16, 36–7, 48–9, 51–2, 120–2, 179–82, 241–4, 257–9, 281
republic regimes 30
Riesman, David 230
Roll, Eric 141
Roman Empire 18–19, 22–23, 26, 33–5, 63, 66, 130–1, 277
Ross, Ian Simpson 78
Royer-Collard 165

Saint-Simon 15
Saints of the Atlas 251
  see also Gellner, Ernest
slavery 118–19
Smith, Adam 51, 73–89, 173, 189, 200, 243, 260, 265, 266, 277, 279–80, 282, 286, 289, 293
  growth and stasis 90–106
  predation to production 123–49
  wealth and liberty 107–22
social unit size 43–5
Sorel, Albert 15, 23–4, 26, 48, 50–1, 58–9, 67
sovereignty 108–10
Spirit of the Laws 13, 19, 21, 23, 48, 58–9, 85, 87, 162
  see also Montesquieu, Baron de
stasis and growth 90–106
Steuart, Sir James 87
Stewart, Dugald 77, 79–80, 85, 90, 112, 141
structural thinking 23–4

Tacitus 62, 64

taxes 54–5, 115–16
Theory of Moral Sentiments 73
  see also Smith, Adam
Tocqueville, Alexis de 153–66, 252, 265, 286, 288, 289, 291–2, 293
  on America 167–92
Tocqueville in England 194
  see also Drescher, Seymour; Tocqueville, Alexis de
towns, growth of 107–12, 277
Townsend, Charles 79
trade between nations 37–41, 111–12
Tucker, Josiah 87

United States see America

violence, ending of 123–5, 127–32
  see also military power

wars 31–35, 218–19, 220
Watt, James 77, 86
wealth 98–106, 107–22, 123–49, 170–6, 197–202, 256
  liberty, equality and 221–47, 286–7
Wealth of Nations 74, 75, 76, 90, 143
  see also Smith, Adam
Weber, Max 182, 255
William the Conqueror 236
Wrigley, E.A. 143

yeomanry 113–14